THE STRUCTURE AND ACTION OF PROTEINS

THE STRUCTURE AND ACTION OF PROTEINS

RICHARD E. DICKERSON AND IRVING GEIS
California Institute of Technology

BENJAMIN/CUMMINGS PUBLISHING COMPANY
Menlo Park, California • Reading, Massachusetts
London • Amsterdam • Don Mills, Ontario • Sydney

to ROBERT B. COREY, *who set an example*

 Printed in the United States of America. Published simultaneously in Canada. Library of Congress Catalog Card No. 69-11112.

ISBN 0-8053-2391-0
GHIJKLMNOP-HA-89876543210

CONTENTS

FOREWORD

To understand the chemical basis of life we must know the machinery that builds up complex plants and animals from simple chemical precursors. What kind of apparatus is it that living cells employ to make large organic molecules in an aqueous medium, at ambient temperatures and in neutral solution, when chemists setting about the same task would employ powerful solvents, high temperatures, low pressures, and strong acids or bases? The answer lies in the efficient catalysts used by living cells to speed up reactions that would normally proceed imperceptibly slowly, ten or even a hundred thousand times. In order to accomplish this, cells provide a special catalyst for each small metabolic step. These catalysts are called enzymes.

The variety of enzymes existing in nature is very large. There are several thousand different ones within the cell of a single bacterium. No one really knows how many there are in a mammal, but it may be several million.

What do these enzymes have in common? They all belong to a class of chemical compounds known as proteins, which means that they consist mostly, and often exclusively, of long chains of amino acids. Compared to familiar molecules like sugar or penicillin, they are enormously large and complex. So large, in fact, that not long ago their constitution and three-dimensional architecture seemed beyond the powers of chemistry and physics. The recent development of chemical and x-ray crystallographic methods for solving the structure of enzymes has been one of the triumphs of molecular biology. This book shows us the results to date. Only about a dozen enzymes have been unraveled, which is a very small number compared to the multitude existing in nature, but it is enough to begin to understand how they work.

So far no novel or sophisticated forces have had to be invoked to explain their action. They can be understood in terms of simple electrostatics. Enzymes recognize their specific reactants by having a site that fits them as a lock fits a key, only more so: the lock often carries electric charges or dipoles exactly compensating those of the reactants. This site often lies in a cleft or cavity below the surface of the enzyme, so that the reactants can be taken out of the surrounding water, and inside the enzyme, as it were. There is a profound chemical purpose in this. In water many reactions proceed slowly because of its high dielectric constant: water acts as an insulator which keeps charged molecules apart. The interior of enzymes, on the other hand, is made up of hydrocarbons, which have a low dielectric constant. In this environment, strong electrical forces can be brought to bear on the reactants, causing them to be altered in a small fraction of a second. So enzymes may be regarded as the organic solvents of the living cell.

Enzymes give us a more detailed picture of catalytic action than any other known chemical system. Knowledge of their structure and function promises to provide a deeper insight into the chemical basis of life, with all that this implies for biology and medicine; it may also suggest new approaches to industrial chemistry which should simplify many of the methods now employed.

This book gives beginners in the field a lucid introduction to the structure and function of proteins. Dr. Dickerson's imaginative writing and Mr. Geis's vivid illustrations combine to make it a useful and inspiring textbook.

MAX F. PERUTZ

Cambridge
June 1969

CHAPTER ONE
THE RULES OF THE GAME

1.1 THE MAKEUP OF LIVING ORGANISMS

A century ago, if a chemist had been asked the difference between the chemistry of living and of nonliving matter, he still might have invoked "life forces," "vis vitalis," or some special properties peculiar to living matter. Nowadays the reply would be given in terms of complexity, organization, and sophistication of the reactions. The chemistry is no different in principle from that of nonliving matter. It is simply more intricate, more subtle, and by the same token more challenging, in the sense that a watch is more interesting than a slingshot and a digital computer than a watch.

The variety of chemical substances present in living organisms is staggering. The most prevalent substance is water, which may represent as much as 65 percent of the weight of a typical mammal. The saline composition of this water reflects the origin of the first living things as tiny enclosed bits of seawater, in which especially efficient series of chemical reactions could evolve with time into metabolisms. Dissolved in this water are a great number of ions and small inorganic and organic molecules: K^+ and Na^+, chlorides and phosphates, organic bases, vitamins, and cofactors. If the living organism is thought of as a complex factory, then these small molecules serve as the nuts, bolts, and cogs that keep the wheels of the factory turning, although many of them are present only in very small amounts. The steel girders of the factory are the inert framework materials: bones, teeth, and the polysaccharide chitin of insects. Lipids—fats, oils, and their derivatives—provide part of the lath and plaster of the walls within the factory. The specifications for everything that goes on in the factory, including the construction of the factory itself, are written down in the nucleic acids: deoxyribonucleic acid (DNA) and ribonucleic acid (RNA). But in many respects the most remarkable chemical substances within living organisms are the proteins.

Proteins are remarkable because they play two distinct and separate roles: as structural materials and as machines that operate on the molecular level. If bones and chitin are the girders, then structural proteins are the bricks of the factory walls, and they provide the other half of the lath and plaster of the interior walls as well. Some of these structural, fibrous proteins are protective: the α and β keratins of skin, hair, silk fibers, nails, and claws. Other fibrous proteins are connective, such as the collagens of tendon. Yet others are motive, such as the actomyosin machinery of muscles.

Proteins can serve as structural material for the same reason that other polymers do—they are long-chain molecules which, by the proper combination of cross-linking, interleaving, and intertwining, can be made to show virtually any desirable bulk properties. This being so, their other class of functions becomes even more remarkable. Proteins also provide the *catalysts* for the reactions in living systems, in the form of relatively small, globular enzyme molecules. Like any other catalysts, these enzymes do not change the point of equilibrium of a chemical reaction, only the rate of getting there. Still, a living system is not at equilibrium until it dies, and in a competition between several possible reactions, the one that goes most rapidly predominates. These macromolecular catalysts therefore control the pathways and the timing of the complicated set of reactions that keeps the organism going.

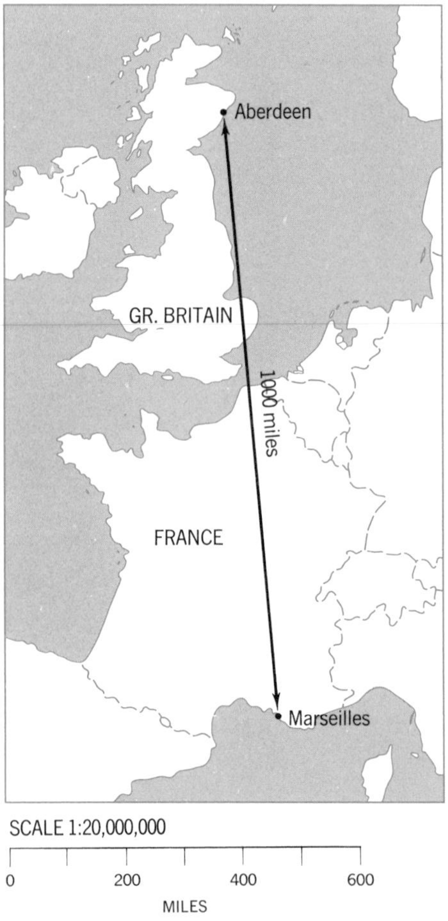

Globular proteins can be relatively small.* Ribonuclease, an enzyme that digests unneeded RNA, is a roughly egg-shaped molecule with a molecular weight of 13,700. Cytochrome c, which transports electrons, is about 25 by 25 by 32 angstroms (Å) and has a molecular weight of 12,400. On the other hand, the gamma globulins, which serve as antibodies, have molecular weights of about 160,000; the enzyme catalase has a molecular weight of 248,000; and the very important enzyme acetylcholinesterase, several million. In cases where the facts are known, these very large globular proteins have been found to be built up from a number of smaller subunits.

How is it that the same substance can be used for both the walls of the reactor and the catalysts within? What is it that gives proteins such a wide range of properties? To answer these questions, we must look at what proteins are and how they are put together.

Aberdeen
GR. BRITAIN
1000 miles
FRANCE
Marseilles
SCALE 1:20,000,000
0 200 400 600
MILES

---

* The molecules on the opposite page have all been enlarged by a linear factor of 20 million times. In order to appreciate what this means, imagine that you were enlarged by this same factor. The earth would then appear to you like a 2-foot diameter sphere, and your index finger would reach from Marseilles to Aberdeen (left).

For such small objects as molecules, miles, feet, and millimeters are all hopelessly large units of length. Molecular dimensions are commonly measured in angstrom units (A), there being 100 million angstrom units to the centimeter. Most molecular bond lengths, then, are between 1 and 3 angstroms.

SCALE 20,000,000:1

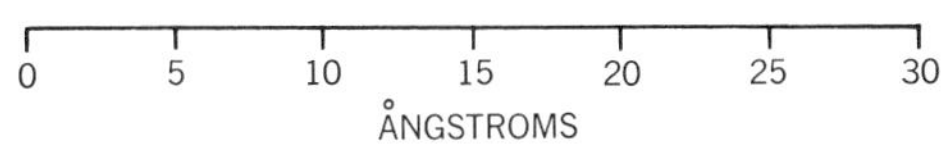
0
5
10
15
20
25
30
ÅNGSTROMS

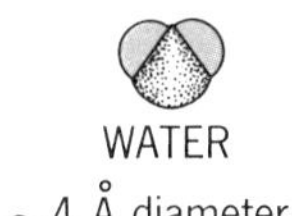
WATER
~ 4 Å diameter

BENZENE
~6 Å diameter

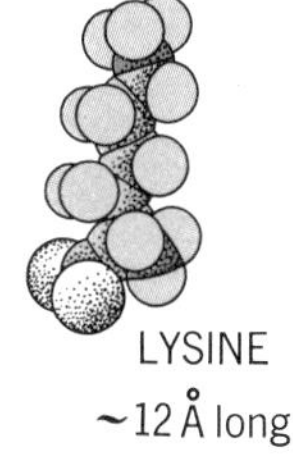
+
LYSINE
~12 Å long

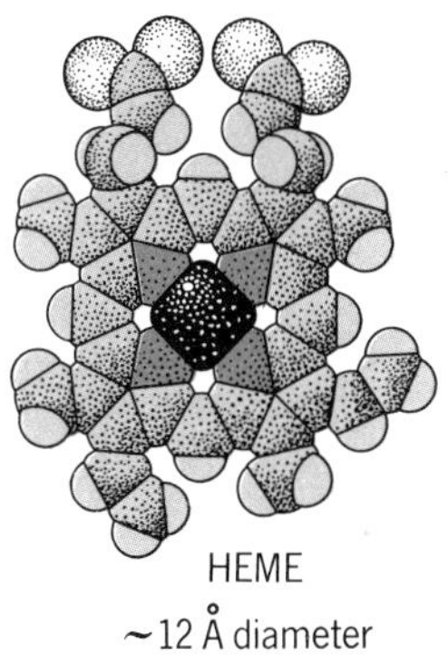
HEME
~12 Å diameter

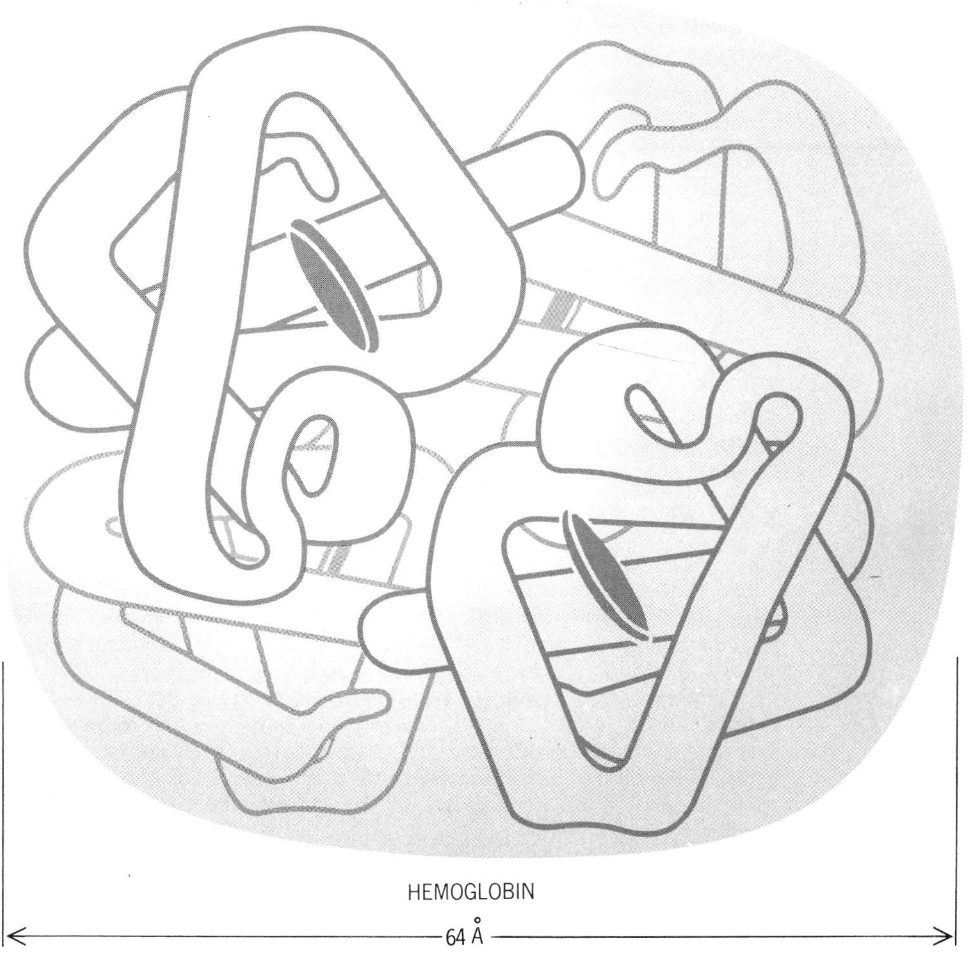
HEMOGLOBIN
64 Å

(3)

1.2 PROTEINS AS POLYMERS OF AMINO ACIDS

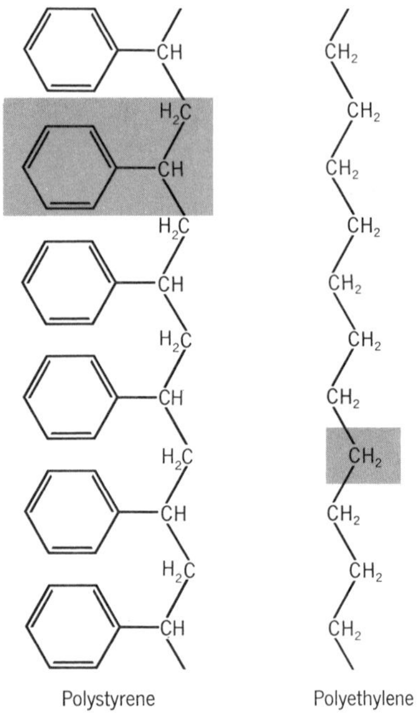

Polystyrene Polyethylene

One repeating monomer chain unit is colored.

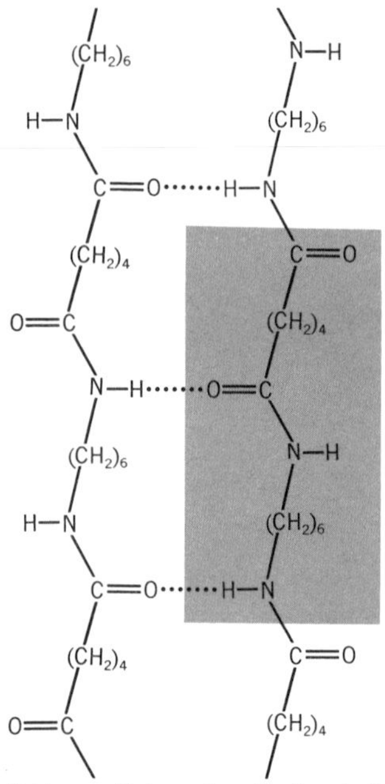

Polyhexamethyleneadipamide (or nylon)

Note the hydrogen-bond cross-linking between chains.

A protein is built up from a long-chain polymer of amino acids, called a polyamino acid or polypeptide chain. Polymers per se can be pretty dull. Polyethylene (left) is good for inert, laboratory beakers and very little else. Polystyrene and nylon have invaluable but limited properties. Branched-chain polymers such as polyurethanes, Bakelite resins, and melamac can form three-dimensional networks with widely varying physical properties. But with all these polymers the tendency is toward inertness rather than catalytic activity.

Polyaminoacids or polypeptides are more versatile because of the great number of different side chains that may be present. The way in which a polypeptide chain is built up by forming peptide bonds between amino acids is shown on the opposite page. Each monomer unit has a side chain, which is usually one of the 20 common types shown on pages 16 and 17. Some, such as valine, leucine, and isoleucine, are hydrocarbons. Aspartic and glutamic acid side chains themselves contain an acidic group; lysine, arginine, and histidine are basic. Some permit cross-linking of chains, either by covalent or hydrogen bonds. It is the variety of possible side groups that makes proteins so useful. If there are 61 units in a chain, all alike, then only one chemical substance is possible. But if each unit has 20 alternatives, the number of possible substances rises to 20^{61}, or 5×10^{79}. As an exercise in exponents, this corresponds to approximately six potential structures for every atom in the universe.* With this range of flexibility, the versatility of proteins is easy to understand.

Proteins are linear polymers, often cross-linked but never branched. In this respect they are like polyethylene and unlike Bakelite. The opposite page shows the two most common types of cross-linking, a covalent disulfide bridge with a bond strength of the order of 50 kilocalories per mole, and a weaker hydrogen bond with about 6 kcal/mole. With $-COOH$ or $-NH_2$ groups present on acidic or

* According to Dauvillier (Les Hypothèses cosmogoniques, Masson, Paris, 1963, p. 67), Einstein's theory of the structure of the universe predicts a total mass of 2×10^{55} grams (g). This would correspond to some 500 billion galaxies, each of about 20 billion solar masses (2×10^{33} g). Of every 1000 atoms in the universe, 875 are hydrogen, 124 are helium, and 1 is O, C, or Ne, other atoms being present in negligible amounts (I.S. Shklovskii and Carl Sagan, *Intelligent Life in the Universe*, Holden-Day, San Francisco, 1966, p. 57). With these data, if n_T = the total number of atoms in the universe and N is Avogadro's number,

$$\frac{0.875 n_T}{N} \times 1 + \frac{0.124 n_T}{N} \times 4 = 2 \times 10^{55}$$

$$n_T = 0.88 \times 10^{79} \text{ atoms}$$

UN-IONIZED AMINO ACID

ZWITTERION

The amino acids exist in solution as doubly charged zwitterions.

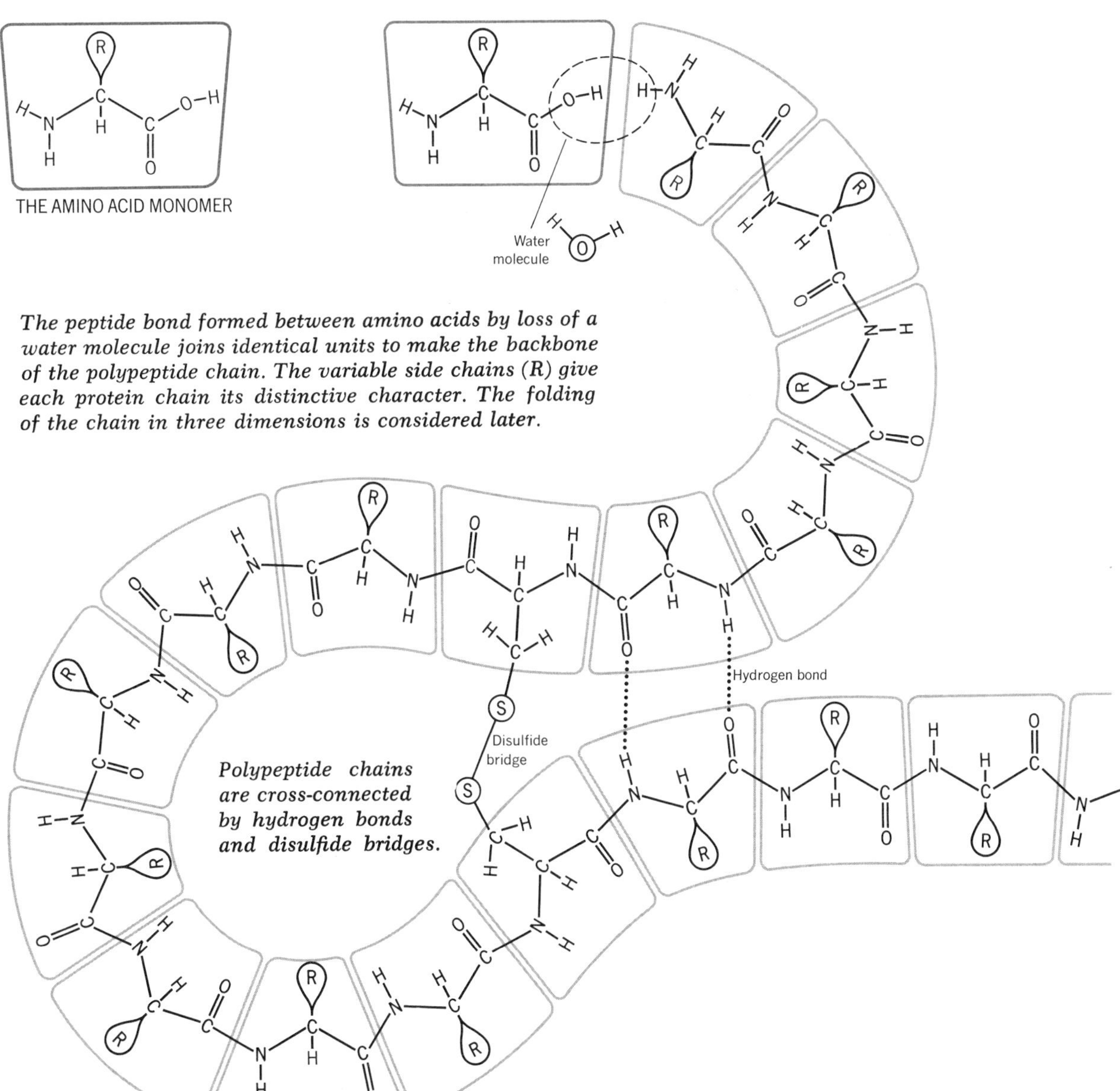

The peptide bond formed between amino acids by loss of a water molecule joins identical units to make the backbone of the polypeptide chain. The variable side chains (R) give each protein chain its distinctive character. The folding of the chain in three dimensions is considered later.

Polypeptide chains are cross-connected by hydrogen bonds and disulfide bridges.

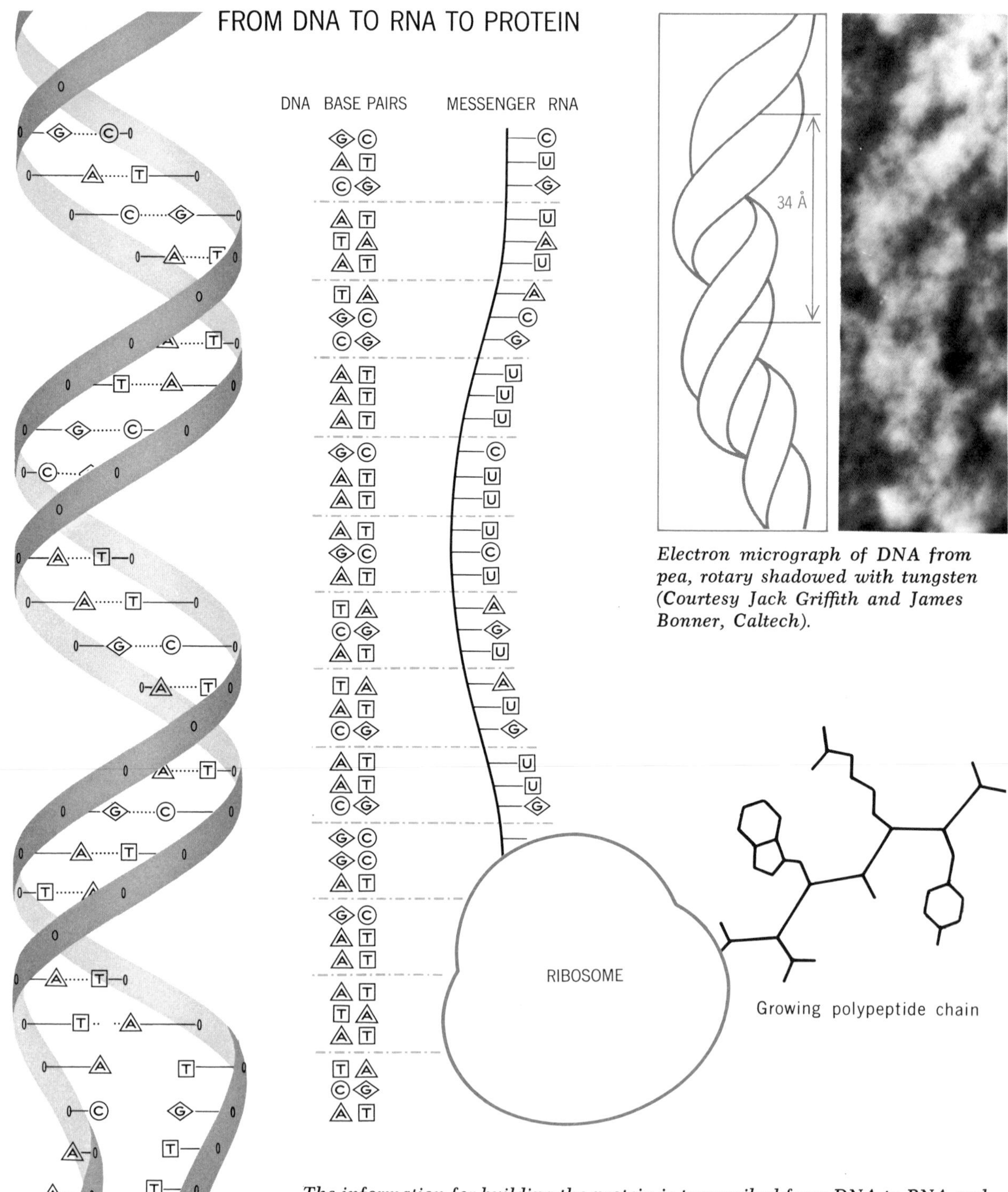

Electron micrograph of DNA from pea, rotary shadowed with tungsten (Courtesy Jack Griffith and James Bonner, Caltech).

The information for building the protein is transcribed from DNA to RNA and translated into the amino-acid sequence at the ribosome. A diagram of the DNA double helix is shown to the left, with an electron micrograph of DNA at 7.3 million times enlargement at the top right.

basic side chains, it might be expected that proteins could be branched-chain polymers of the type shown to the right. In fact they never are, and the reason lies in the way in which proteins are made.

The information on how to make a particular protein is stored in the nucleus of the cell as a coded sequence of organic bases along a molecule of DNA (or RNA in some simple organisms). Four bases are possible: adenine (A), thymine (T), guanine (G), and cytosine (C). Three successive bases along the DNA code for one of the 20 common amino acids. The DNA itself is retained as "archive" reference material; whenever a given protein is to be synthesized, a copy of the information is made on RNA. The "messenger" RNA diffuses out of the nucleus to a small cellular body called a ribosome, where the protein is synthesized from amino acids and where the sequence of triplets of bases originally present in DNA is translated into a sequence of amino acid side chains along the polypeptide. For the fascinating details of this process, which are only hinted at here, the reader is referred to James D. Watson's *The Molecular Biology of the Gene*. The important point here is that the original information storage device is linear. A sequential storage device can use a much simpler readout system than a two- or three-dimensional matrix store. This is why tape recorders use tape and not sheets or blocks of magnetic material. But it is difficult to see how the instructions for branched-chain polymers could conveniently be coded into a straight-chain polynucleotide. If three-dimensional structures are needed, a more efficient method might be to produce linear polymers, and then to provide spontaneous mechanisms for cross-linking and folding the chain by means of a proper sequence of the right kinds of side chains. This appears to be the case for proteins. Hydrogen bonds and disulfide bridges act as specific folding influences. The tendency of hydrocarbon-like side chains to segregate together in an aqueous environment acts as a nonspecific influence, but one stabilizing in free energy.

There appears to be no spontaneous formation of a polypeptide bond after synthesis of the chain at the ribosome. There are no enzymes present in living systems that branch-link chains as shown to the right. To date there is not even any clear evidence of head to tail linking of two polypeptide chains by a peptide bond after synthesis to form a longer straight chain. The reason is easy; such enzymes would be too dangerous. Enzymes, like all catalysts, speed up forward and reverse reactions equally and accelerate only the drive toward equilibrium. Polymerases are also depolymerases. An enzyme that would link two amino acids to form a peptide bond would just as readily break one already formed. With the immense number of peptide bonds present in the protein of an organism, simple mass-action arguments will show that such an enzyme would spend most of its working life as a *de*polymerase and would work havoc on its host. With the sequence of side chains critical in deter-

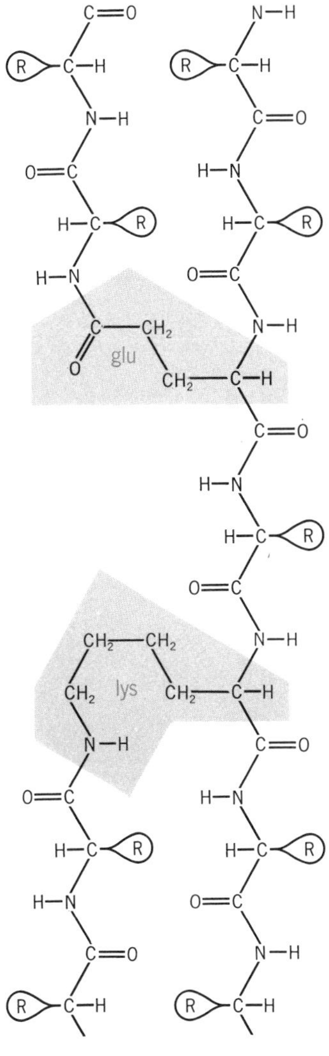

Side-chain branching of a type that is never found in proteins. Top, formation of a peptide bond with the carboxyl group of a glutamic acid side chain (glu); bottom, same linkage with the amide group of lysine (lys).

mining the properties of a protein, the danger of scrambling is too great for a peptide polymerase to be tolerated. Only in restricted localities where they have a digestive function are such polypeptide depolymerases, such as trypsin, chymotrypsin, or pepsin, found.

The current picture (would "dogma" be better?) is that protein is synthesized as a linear polymer and that its subsequent quite specific folding is determined entirely by the kind and distribution of side chains. The process is quite efficient, but if anything can be said to have come out of three or more billion years of evolution, it is efficiency.

A substituted hydrocarbon "pseudopolypeptide" chain. Its dimensions are similar to a protein chain, but its chemical properties are quite different.

1.3 THE BACKBONE OF THE POLYMER

Many of the special properties of a polypeptide chain arise from the nature of its backbone chain. Its distinctive feature is the group —CO—NH—, called a peptide bond or amide link. A polymer with "pseudopeptide" bonds as at the left would have similar overall dimensions, but its chemical behavior would not even remotely resemble that of a protein.

The CO and NH groups are capable of forming cross-links between chains and of overcoming the absence of true branching when building up three-dimensional structures. Several such common

Approximate electron density distributions for electrons in atomic orbitals.

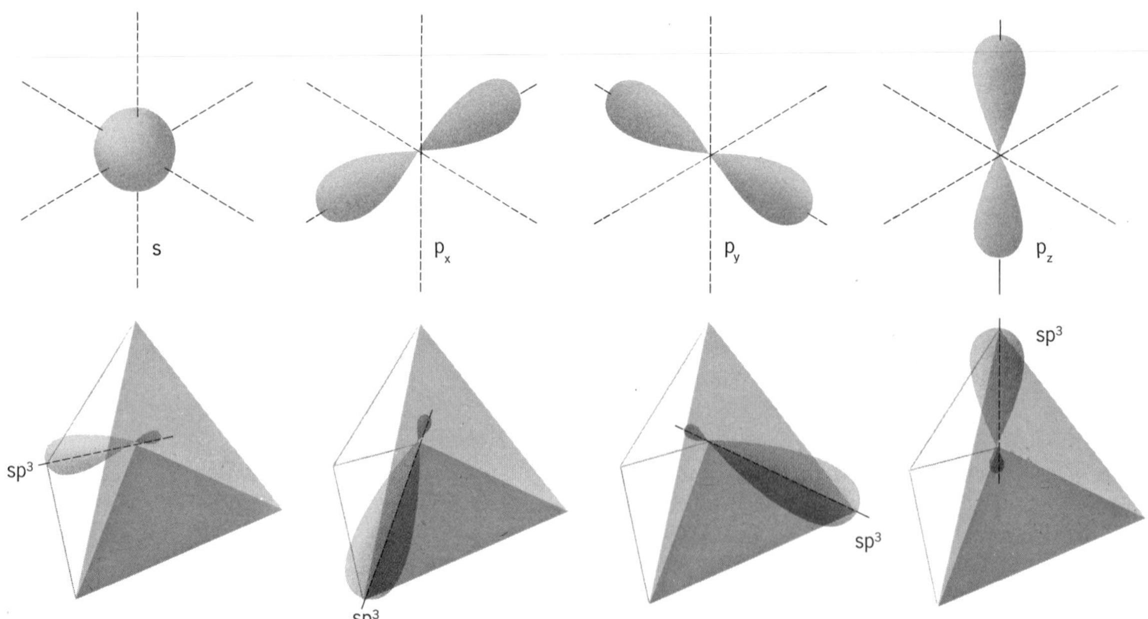

The four tetrahedrally oriented sp^3 hybrids obtained from the four orbitals above.

structures are discussed in Chapter 2. The peptide bond also severely limits the ways in which the chain can fold: all four atoms in the —CO—NH— group have to lie in the same plane. The easiest way to see why this is so is to look at a simple molecular orbital explanation of chemical bonds.

Carbon, nitrogen, and oxygen atoms, when they form bonds, ordinarily use their $2s$, $2p_x$, $2p_y$, and $2p_z$ atomic orbitals. In addition to the two electrons in the filled $1s$ orbital, which play no part in bonding, carbon has four more valence electrons, nitrogen has five, and oxygen has six. Hydrogen has only one electron, half filling its spherical $1s$ orbital.

In methane, CH_4, the four carbon $2s$ and $2p$ orbitals do not combine directly with the hydrogen $1s$, for the hydrogen atoms are observed to be tetrahedrally arranged about the carbon. The carbon orbitals may be thought of as being combined (hybridized) to form four equivalent sp^3 atomic orbitals, directed to the corners of a tetrahedron. These then each combine with one $1s$ hydrogen orbital to form a molecular orbital that, when filled by two electrons, builds one C—H bond. The eight valence electrons are thus accounted for in four molecular orbitals. Because each C—H bond electron cloud is cylindrically symmetrical about the bond axis, these orbitals are referred to as σ-type molecular orbitals. The extra stability of the methane molecule over its five isolated atoms can be expressed as that of four C—H bonds with bond energies of 99 kcal/mole each.

The sp^3 *hybrid atomic orbitals should be thought of as hypothetical intermediates leading to the formation of four single bonds upon interaction with orbitals from neighboring atoms. The small price paid in instability in forming the* sp^3 *hybrids is more than compensated for by the extra stability of the four final bonds.*

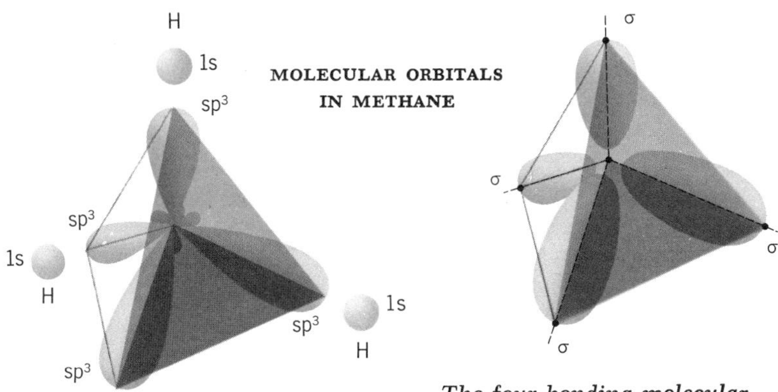

Formal representation of the four carbon sp^3 *and four hydrogen* $1s$ *atomic orbitals which enter into the binding.*

The four bonding molecular orbitals formed from the atomic orbitals to the left. Each molecular orbital is occupied by a pair of electrons. The "σ" indicates that the orbital is cylindrically symmetrical about its C—H bond axis.

The bonding in ethane, CH_3-CH_3, is similar. There are 14 valence electrons, 4 each from the carbons and 1 each from the hydrogens. There are six σ-type C–H molecular orbitals and one σ-type C–C orbital (see below). Each such orbital is occupied by one pair of valence electrons. The C–H bond energy is again 99 kcal/mole and that of the C–C bond is 83 kcal.

Ethylene, $CH_2=CH_2$ (shown on the facing page), illustrates the molecular orbital picture of a double bond. In this molecule, on each carbon atom, the 2*s* and two of the 2*p* orbitals are hybridized to form three equivalent sp^2 orbitals, lying 120° apart in a plane. The unused 2*p* orbital has its axis perpendicular to this plane, and its electron cloud is symmetrical above and below the plane. Four σ-type C–H bonds are formed, in the usual way, and one σ-type C–C bond. These are occupied by 10 of the 12 valence electrons. But in addition, the two unused 2*p* orbitals also combine to form a different type of C–C molecular orbital. This orbital is not cylindrically symmetrical about the C–C axis, being made up of two lobes of electron density above and below the plane of the molecule, and reflecting the mirror symmetry of the atomic orbitals from which it arose. Such an orbital is called a π-type orbital. The last two electrons fill this π orbital to form the second carbon–carbon bond.

The asymmetry of the π orbital requires that all six atoms in the ethylene molecule lie in one plane, for a twist about the C–C bond would forcibly uncouple the two 2*p* orbitals that form it and reduce the double bond to a σ-type single bond. The measured bond energy of a carbon–carbon double bond is 147 kcal/mole; that of a single

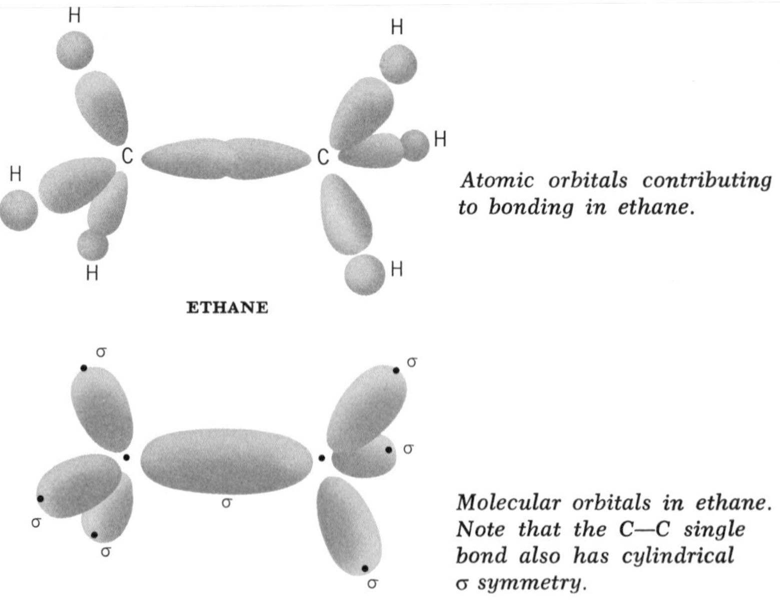

Atomic orbitals contributing to bonding in ethane.

Molecular orbitals in ethane. Note that the C—C single bond also has cylindrical σ symmetry.

bond is 83 kcal. Hence an extra 64 kcal/mole would be required to twist one end of the ethylene molecule by 90°.

So far, the simple molecules we have mentioned can be described by single or double bonds between pairs of atoms. This is not always so. When a carboxyl group is ionized, for example, it might be supposed that one of the carbon–oxygen bonds would remain a carbonyl double bond, 1.23 Å long, and that the other would remain a single bond, 1.36 Å long, to the now negatively charged oxygen. In fact, crystal-structure analyses of salts of carboxylic acids have shown that the two carbon–oxygen bonds are usually equivalent, with an intermediate length of 1.26 Å. One way of describing this is to say that the true bond structure has the character of a mixture of the two extreme simple bond models, or resonance models. This terminology is deceptive, for the word "resonance" suggests a flipping back and forth between the two structures, which is wrong. It is better to say simply that the single bond–double bond picture is too naive and that the real structure has two equivalent bonds which are more than single and less than double ([c], at right). The negative charge is spread over the entire carboxyl group, and the two bonds each have *partial* double-bond character. The electrons that would have been confined to the region of one double bond are said to be "delocalized." It is generally true that, other things being equal, an electron with more room in which to move around will have a lower energy. Bond structure ([c], at right) is more stable than either (a) or (b) by a "resonance stabilization energy" of 28 kcal/mole.

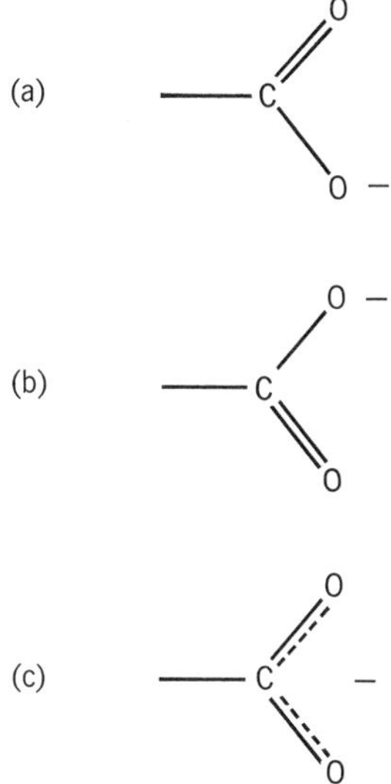

Resonance models in an ionized carboxyl group. Drawings a and b show extreme forms of model bond structures; c, the observed structure.

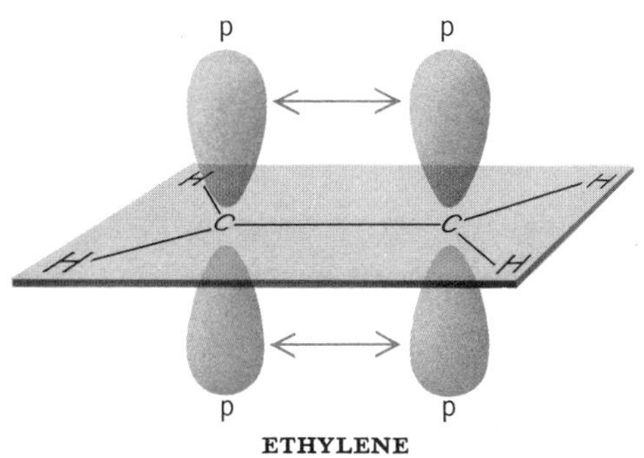

The 2p *atomic orbitals used in the second bond of a C=C double bond.*

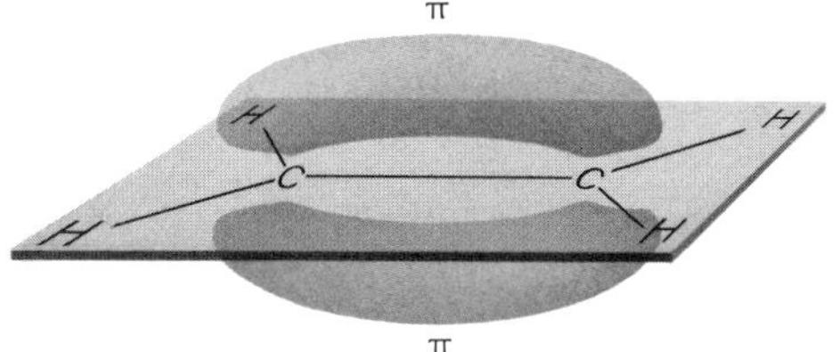

A closer approximation to the actual electron cloud of the second bond. Note that the cloud is not *symmetrical with rotation about the C—C axis. A bond with this symmetry is called a π-type bond.*

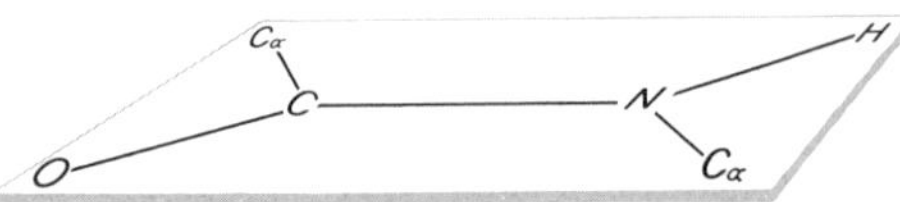

The link between carbonyl carbon and nitrogen (—CO—NH—) is the fundamental structural unit of the polypeptide chain.

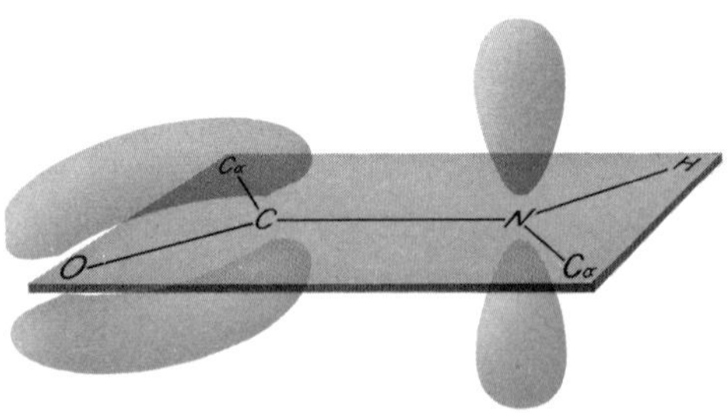

A pure double bond between C and O would permit free rotation around the C—N bond.

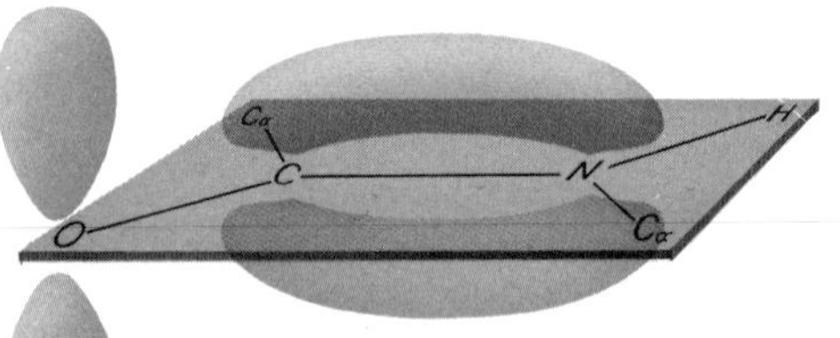

The other extreme would prohibit C—N bond rotation but would place too great a charge on O and N.

All of this is by way of an introduction to the peptide or amide bond. Again, two extreme resonance bond models can be drawn, and the truth is something in between. One extreme (center left) has a C=O double bond and a C—N single bond, with the unused electron pair on the nitrogen occupying its unhybridized $2p$ orbital. (Both C and N here use sp^2 hybridization.) At the other extreme (lower left), the double bond is between carbon and nitrogen, with a single C—O bond and a $2p$ lone pair on oxygen. Because of this shift of electrons from nitrogen toward oxygen, there is a net positive charge at N and a negative charge at O in this structure.

The true situation is shown at the bottom of the page. The carbon–oxygen bond is less than double, and the carbon–nitrogen bond has partial double-bond character. Electrons are delocalized and occupy a π-like molecular orbital that extends over all three atoms. Thermodynamic measurements on simple amide molecules show that the extra resonance stabilization energy of the true structure over the C=O, C—N extreme is around 21 kcal. Hence energy is required to twist the amide link about the C—N axis, and a twist of $\theta°$ will require an energy of $E=21 \sin^2 \theta$ kcal/mole.

There are no examples known in which the amide bond is not planar, except for small twists in one or two crystal forms induced by intermolecular forces. There is no reason to believe that in solution or in vivo, amide bonds are anything other than planar. One other possibility remains, however. The peptide bond as drawn below has α carbons (the carbons that bear the side groups) at diagonally opposite corners of the amide plane, in what is called the "*trans*" configuration. One end could be rotated 180° to bring the α carbons closer together without destroying the planarity. But this "*cis*" con-

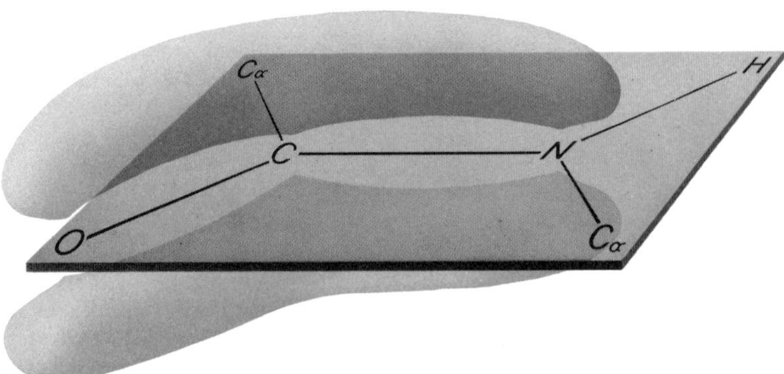

The true electron density is intermediate. The barrier to C—N bond rotation of about 21 kcal/mole is enough to keep the amide group planar.

figuration is slightly less favorable, probably because the bend that it gives to the polypeptide chain causes steric problems. The *cis* configuration has been found only in a few isolated polypeptides and only occasionally in Pro residues in proteins.

One of the most useful references for a commercial architect is his architectural standards data book, containing the dimensions of commercially available building components: masonry, hardware, window and door casings, and other fittings. The architect who overestimates the size of his bricks by 5 percent, or the carpenter who thinks that a two-by-four is 2 in. by 4 in., will be in trouble. In a similar way, if we want to learn how proteins are put together, we must first find out the standard dimensions of a polypeptide chain—in this case, bond lengths and angles.

In the late 1930s, Linus Pauling, Robert Corey, and their colleagues at Caltech set out on a systematic x-ray diffraction study of crystals of amino acids, dipeptides, and tripeptides to find these standard values. In the absence of any ground rules, the folding schemes proposed by various people for polypeptide chains in proteins were often pretty wild. It had become apparent from x-ray diffraction patterns that many fibrous proteins were helical, and various helices were proposed with two, three, four, five and more amino acid residues per turn of helix. None was completely satisfactory. In 1951 the Caltech group summarized their findings in the consistent set of dimensions for a polypeptide chain (1, 2, 3),* similar to the most recent values shown below and to the right.

* References are grouped by chapters beginning on page 115 and are noted in the text by numbers in parentheses.

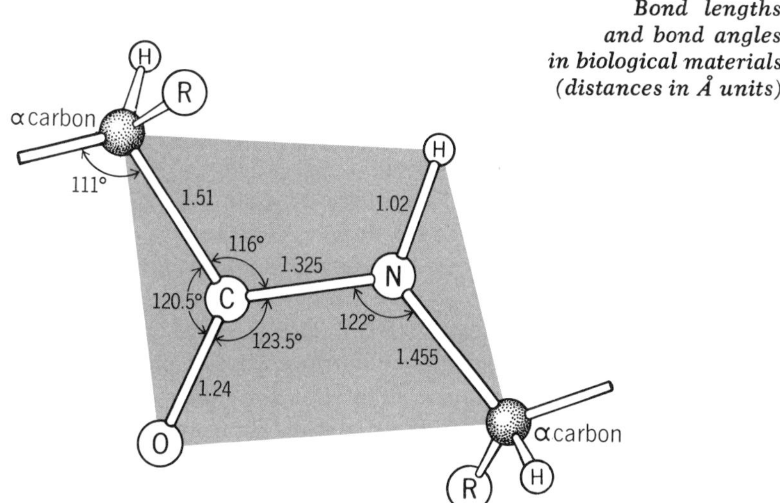

Basic dimensions of the peptide bond.

Bond lengths and bond angles in biological materials (distances in Å units).

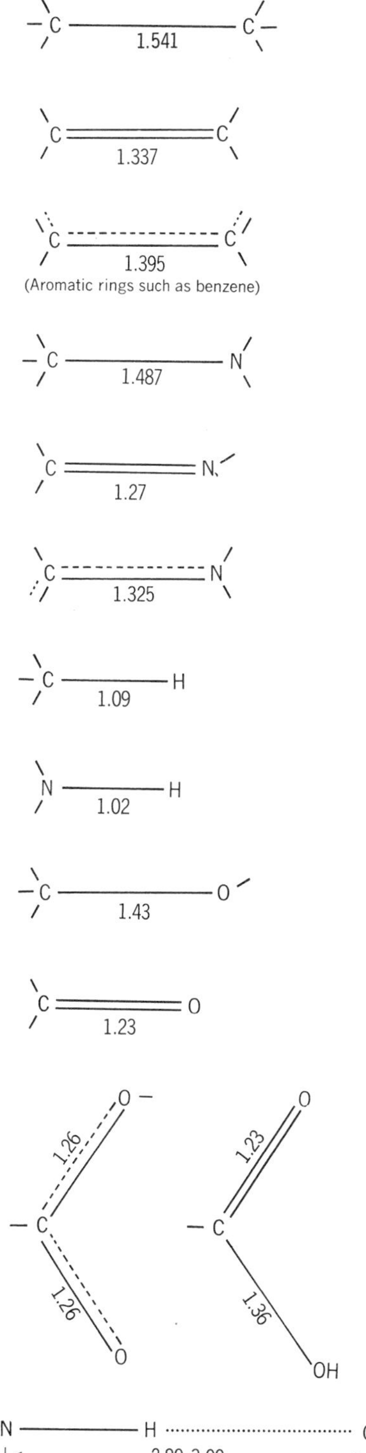

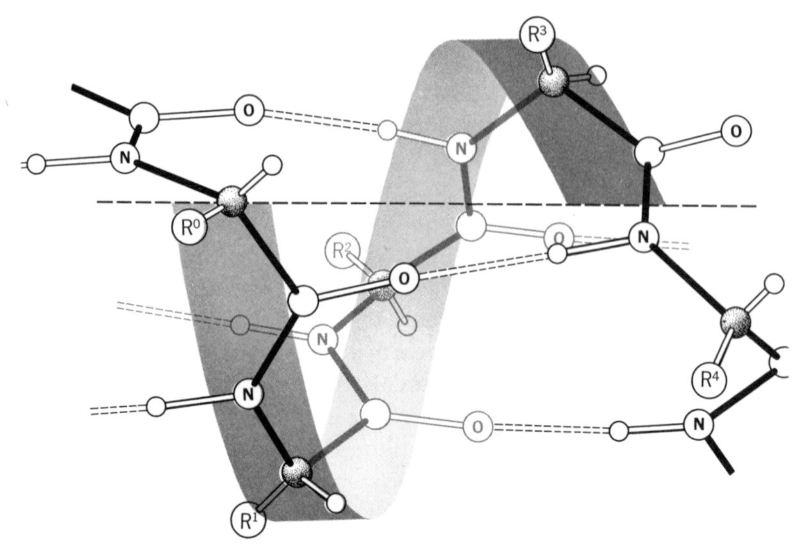

In this α helix, each side chain (R^0, R^1, R^2, R^3, . . .) marks one repeating unit of the helix. Note that one complete turn takes more than three units but less than four (3.6, to be exact).

Pauling and Corey also proposed a helical structure that had come from careful model building with their new parameters, the α helix with 3.6 residues per turn (above). Literally within hours after reading the first report of the Caltech work, M. F. Perutz of Cambridge had proved the presence of the α helix in hair keratin and other proteins by finding a previously overlooked but critical x-ray reflection (4), and it rapidly became obvious that the α helix was one of the fundamental ways of folding a polypeptide chain. Crystallographers are accustomed to twofold, threefold, fourfold, sixfold, and other screw axes or helices with an integral number of units per turn. It apparently never occurred to anyone that in a polypeptide chain, the units per turn could just as well be irrational as integral. The α helix, like so many good ideas, was unthinkable beforehand but self-evident afterward.†

Any carbon atom with four different groups bonded tetrahedrally to it can have them bonded in two different ways related by a mirror reflection. The two structures will rotate polarized light in opposite senses and have come to be called D and L optical isomers, from dextro (right) and levulo (left). The *absolute* configuration of the bonds about such an asymmetric carbon atom cannot be learned from its optical rotation alone. But x-ray diffraction studies have led to the absolute handedness of a great number of asymmetric compounds, including amino acids.

† For an account of the way in which ideas about protein structure developed, see reference (5), and for a thorough discussion of polypeptide chain conformation, see reference (6).

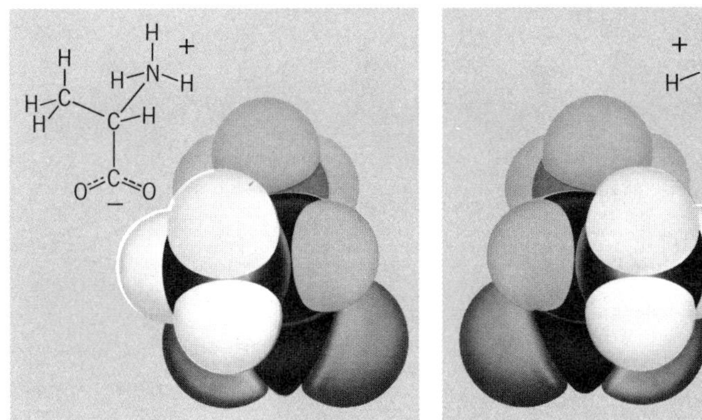

L-alanine and its enantiomorph (mirror image), D-alanine.

Only one optical isomer of amino acids is found in the higher organisms, the form shown above, called an L-amino acid. The symbols L and D have come to have only formal significance and do not themselves refer to rotation of light to the left or right, respectively. One device to fix the absolute configuration in your mind is to imagine that you are crossing a humpbacked bridge at the α carbon, in the direction CO to NH. (Remember the sense as "ONward.") With an L-amino acid the side chain at the summit of the bridge is to the left; with a D-amino acid it is to the right.

The problem of the origin of this preference for just the L form is tied up with that of the origin and evolution of life itself. Some have looked for physical forces that would make the L form preferable: magnetic fields, preferred polarization of ultraviolet light, or even Coriolis forces from the rotation of the earth. Others have attributed it to chance; when life finally developed at one place on the earth, it was so much more efficient than less organized chemical systems that it swept the planet instantaneously as cosmic processes go,* eliminating the opportunity for life to re-evolve somewhere else with active compounds of the opposite symmetry. The question is still open.

---

* Meaning perhaps tens of thousands of years.

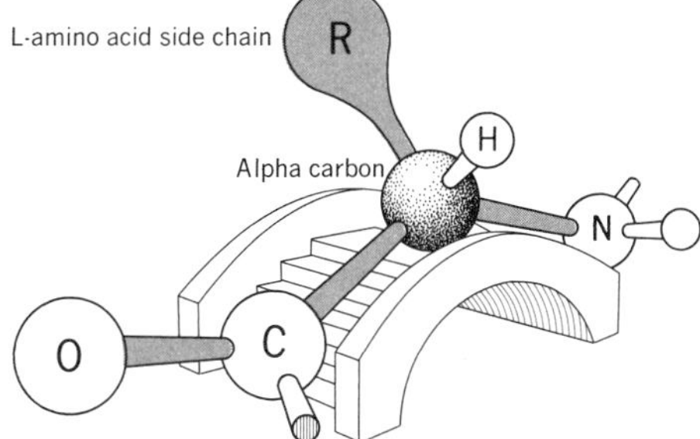

When crossing the bridge as shown, remember "L- to the left; D- to the right".

main chain
side chain

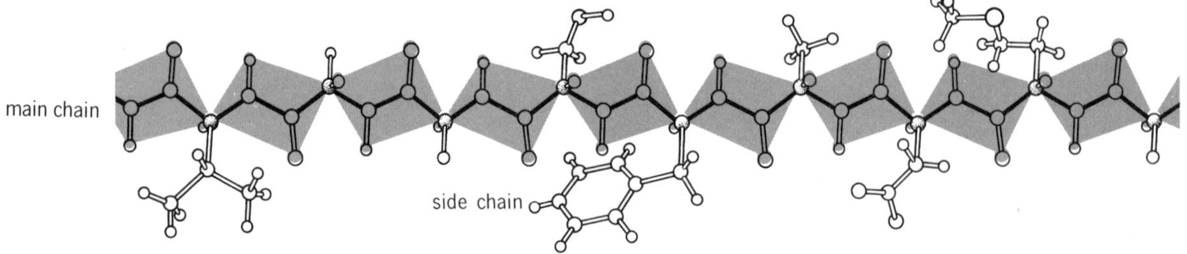

The amide link (CO—NH) is the repeating unit of the main chain; the side chains vary.

POLAR RESIDUES

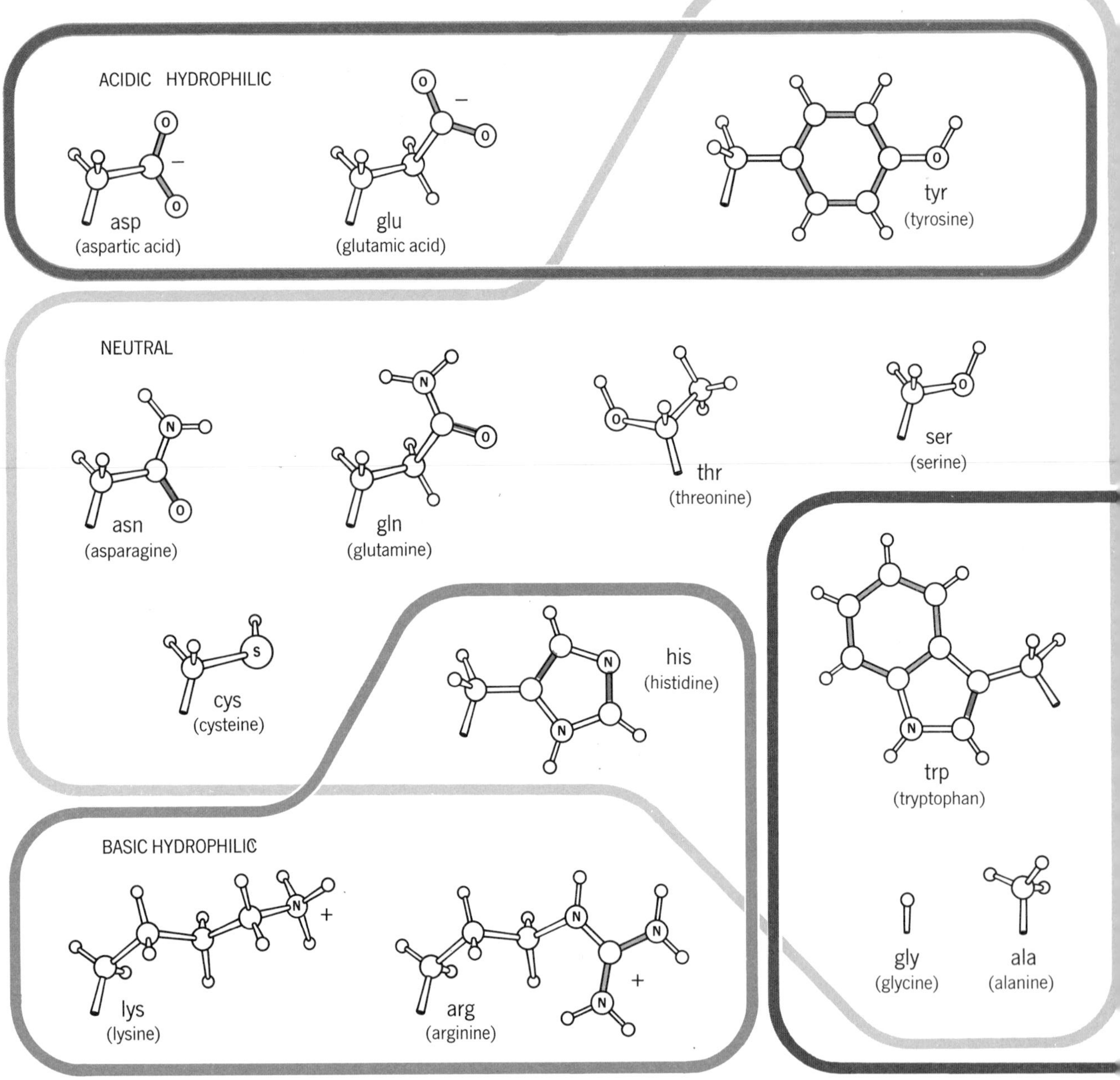
ACIDIC HYDROPHILIC
asp
(aspartic acid)
glu
(glutamic acid)
tyr
(tyrosine)
NEUTRAL
asn
(asparagine)
gln
(glutamine)
thr
(threonine)
ser
(serine)
cys
(cysteine)
his
(histidine)
trp
(tryptophan)
gly
(glycine)
ala
(alanine)
BASIC HYDROPHILIC
lys
(lysine)
arg
(arginine)

If the polypeptide chain provides the fundamental pattern, the ground bass of the composition, it is the side chains that build the melody.* Out of the chemical and x-ray diffraction studies of proteins have gradually come guidelines as to the kinds of amino acid side chains and the parts they play in building a protein.† There are three general categories of side chains: nonpolar, polar but uncharged, and charged polar. On these two pages are shown the 20 common side chains grouped by categories. The nonpolar residues include those with aliphatic hydrocarbon side chains: Gly, Ala, Val, Leu, Ilu, Pro, one aromatic group, Phe, and one "pseudohydrocarbon," Met. (Throughout the book, amino acids will be represented by their three-letter symbols rather than by name.) The polar but neutral category contains two hydroxyl-containing residues, Ser and Thr; two amides, Asn and Gln; two with aromatic rings, Tyr and Trp; and one with a sulfhydryl group, Cys. In the charged polar class are two acidic groups, Asp and Glu; and three bases, His, Lys, and Arg.

* A protein for every purpose, or as the musicologist would put it, "Chaconne à son goût."

† The material of this section is based largely upon the results of crystal-structure analyses of myoglobin, hemoglobin, lysozyme, carboxypeptidase, ribonuclease, and α chymotrypsin, references (7)–(14). For measures of relative hydrophobicity, see references (15) and (16).

NONPOLAR RESIDUES

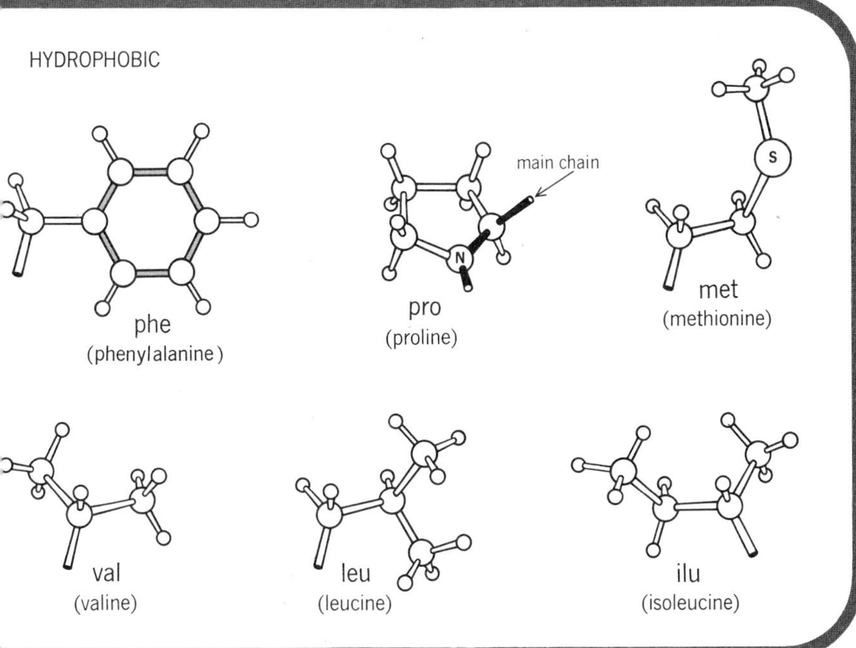

○ *Side-chain carbon*

Ⓝ *Nitrogen*

Ⓞ *Oxygen*

o *Hydrogen*

Ⓢ *Sulfur*

Main chain

Single bond

Double bond

Resonance bond of intermediate character

Polar side chains usually found on the surface of the molecule are on the opposite page. At the left are the side chains usually found in the interior.

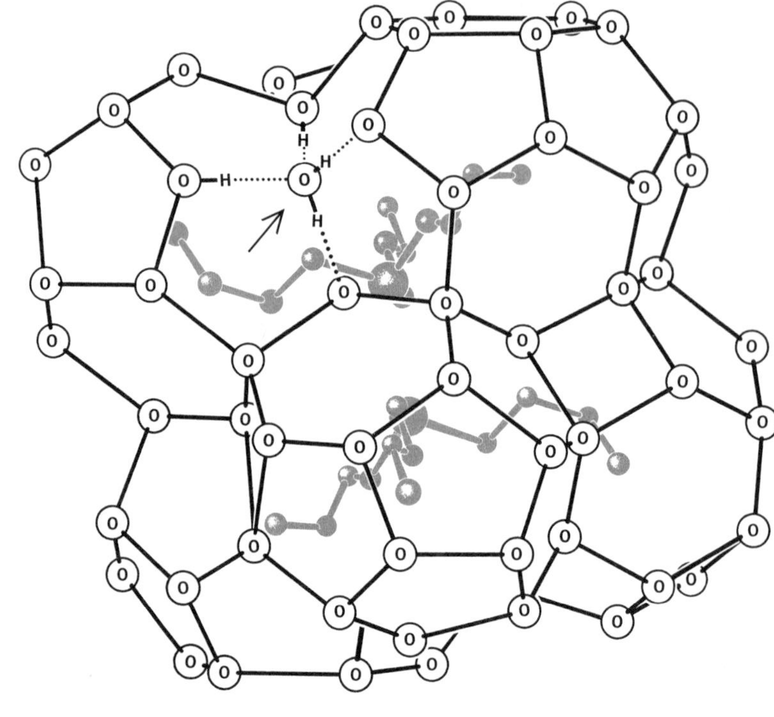

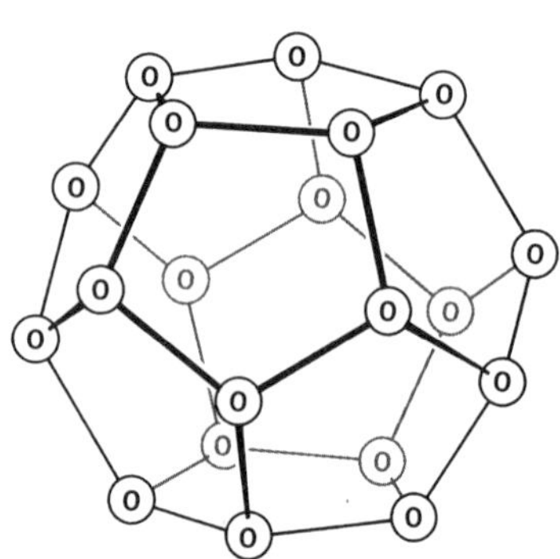

CLATHRATE STRUCTURES

Clathrate structures are ordered cages of water molecules around hydrocarbon chains. The dodecahedral cage (above) of water molecules is a common building block in clathrates. To the right is a portion of the cage structure of $(n\text{-}C_4H_9)_3S^+ F^- \cdot 23\ H_2O$. *The trialkyl sulfur ion nests within the hydrogen-bonded framework of water molecules. In the intact framework, each oxygen is tetrahedrally coordinated to four others. One such oxygen atom and its associated hydrogens are shown by the arrow.*

Proteins have evolved for operation in an aqueous environment. Part of the function of an enzyme is often to provide less polar surroundings for the molecule or molecules on which it acts, its substrate, than are obtainable in solution. The chemistry of an enzyme–substrate interaction is very much affected by the polarity of the solvent. We shall see in Chapter 4 that the reactivity of lysozyme's *own* acidic side chains is changed by their local surroundings, and that this is critical in its catalytic activity.

The nonpolar side chains, then, provide the opportunity for a little nonaqueous chemistry. They also help to hold the molecule together. When a hydrocarbon chain is in an aqueous medium, it forces the neighboring water molecules to form a cage-like or "clathrate" structure in the immediate vicinity, as shown above. This restricts the motion and number of possible arrangements of the water molecules and lowers their entropy. If these hydrocarbon molecules are segregated in one place instead, then the liberated molecules of water are free to adopt a much less ordered arrangement, and the entropy of the solution rises. This is why oil droplets separate out spontaneously in water—the driving factor is entropy more than it is energy. Kauzmann (17) has calculated that for every nonpolar, hydrophobic side chain of a protein that is removed from an aqueous to a nonpolar environment, the protein gains an extra 4 kcal of free energy stabilization, chiefly from this entropy effect. This makes the segregation of hydrophobic side chains a powerful factor in stabilizing a protein molecule in aqueous solution, and leads to the "oil drop" model of a globular protein as a polypeptide chain with all of its nonpolar groups inside and its polar groups outside.

The remainder of the discussion is really concerned with elaborations and exceptions to the "nonpolar in, polar out" rule. The rule applies with force only to the larger nonpolar groups: Val, Leu, Ilu, Pro, and Phe. Gly and Ala are so small that they apparently can be accommodated on the interior or the surface with equal ease. Nonpolar groups *have* been found on the surface where they appear to play a role in binding of subunits in complex proteins or of substrates in enzymes. They do lead to instability, and it is safe to assume that when they are found on the outside, they must be there for a specific purpose. On the other hand, charged side chains are almost never found away from the surface of the molecule and are much more restricted as to environment. Most charged groups that are not specifically involved in the function of the protein seem to exist for the sake of interacting with the solvent and keeping that part of the chain in polar surroundings.

The neutral polar residues are usually outside the molecule, but can be inside if their polar groups are "neutralized" by hydrogen bonding to other like residues or to the carbonyl C=O group of the main chain. Ser, Thr, Asn, and Gln are often used to cross-link two chains by means of hydrogen bonds. Tyr and Trp have been found inside and outside, but when Tyr is inside, its hydroxyl group is always hydrogen-bonded.

The "charged polar" groups, acidic or basic, can exist in either uncharged or charged form, depending on the pH of the surroundings. Under acid conditions, Asp and Glu have an uncharged carboxyl group, whereas His, Lys, and Arg each are protonated and carry a positive charge. Under basic conditions, the carboxyl groups of Asp and Glu will be ionized, and His, Lys, and Arg will be uncharged. The actual ratio of the acidic to the basic form of a given residue depends on its strength as an acid or base. The pH at which the two forms are present in equal amounts is called the pK. Moving one or two pH units away from the pK in either direction causes the ratio of the two forms to change to 10:1 or 100:1, respectively.

Asp and Glu have pK values of 3.86 and 4.25, respectively. His is 6.00, Lys is 10.53, and Arg is 12.48. Under normal physiological conditions, near pH 7, Asp and Glu will be almost entirely in their basic, or charged, form. Lys and Arg will be in their acid form, positively charged, but His will be largely uncharged. It will be about 10 percent protonated and is capable of playing a dual role. The Tyr hydroxyl group is weakly acidic with a pK of 10.07. Only about 0.1 percent will be ionized at pH 7, and hence Tyr has been classed as an uncharged polar group. The forms shown on page 16 are those most prevalent at pH 7.

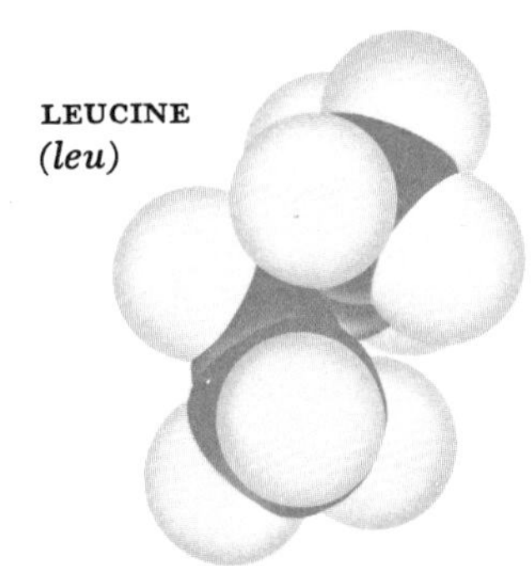

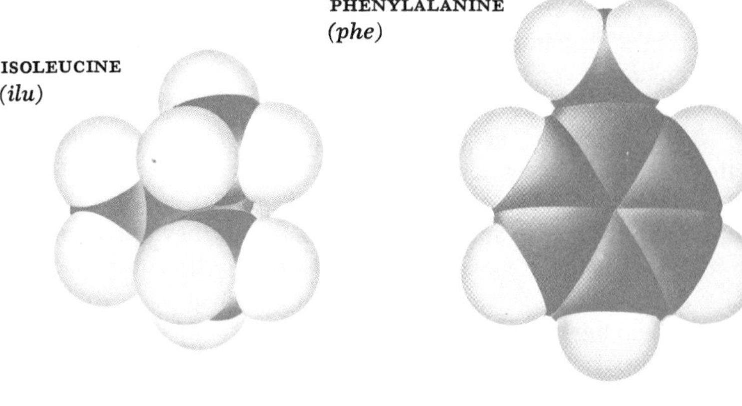

Nonpolar side chains become increasingly hydrophobic from the smallest gly (lower left) to the largest phe (above). These drawings have been made from Corey-Pauling-Koltun (CPK) space-filling models, which give an accurate representation of the bulk of the atoms.

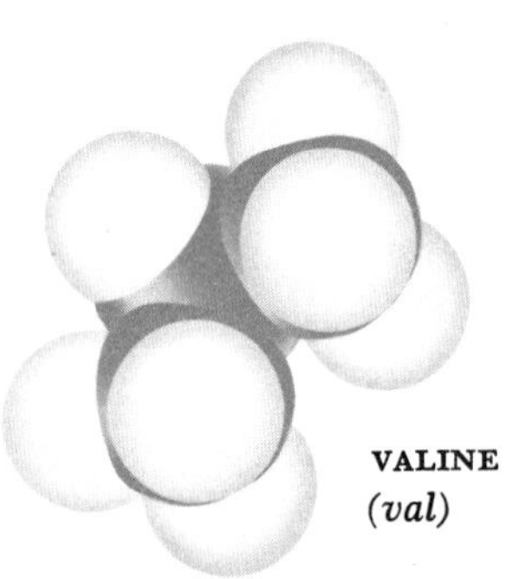

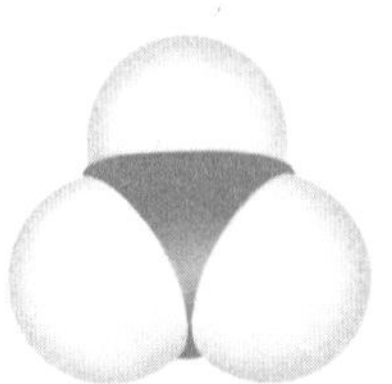

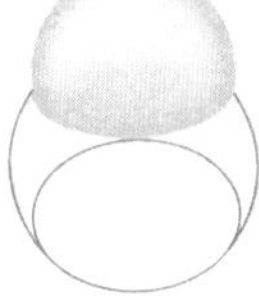

It is an example of the economy of the evolutionary process that most amino acids have more than one role. The hydrocarbons Ala, Val, Leu, and Ilu increase in bulk and vary in shape and, as they increase in bulk, increase correspondingly in hydrophobic character. But if Leu and Ilu are useful, why is there no straight-chain $-C_4H_9$ norleucine? The reason, suggested by E. L. Smith (18), is that Met is available with a similar size and shape but with other chemical properties that norleucine would not have. Met can also use the lone-pair electrons of its sulfur atom to form metal-ligand bonds. It would be disadvantageous to use norleucine instead of Met and inefficient to use them both. The result is that the process of natural selection has left Met as one of the basic 20 amino acids but not norleucine.

Other amino acids similarly have their special properties. In addition to its hydrophobic bulk, Phe has the ability to interact with other aromatic rings by means of overlapping π electron clouds, a property it shares only with Tyr and Trp. This is important in interactions with the heme group in myoglobin and hemoglobin (see page 59), and probably in cytochrome c as well. Ser can do more than serve as a "cross-linking hydrocarbon." In the proteolytic enzymes trypsin and chymotrypsin, one particular Ser at the surface is essential for catalytic activity. The lone electron pair at one ring nitrogen atom in His makes it, like Met, a potential metal ligand, important in heme iron binding in myoglobin, hemoglobin, and cytochrome c. His is also involved in the active sites of trypsin, chymotrypsin, and ribonuclease, among other enzymes.

Cys plays a crucial role in determining the folding of many proteins by virtue of the ability of two such residues on different polypeptide chains to be oxidized to form a disulfide bridge (far

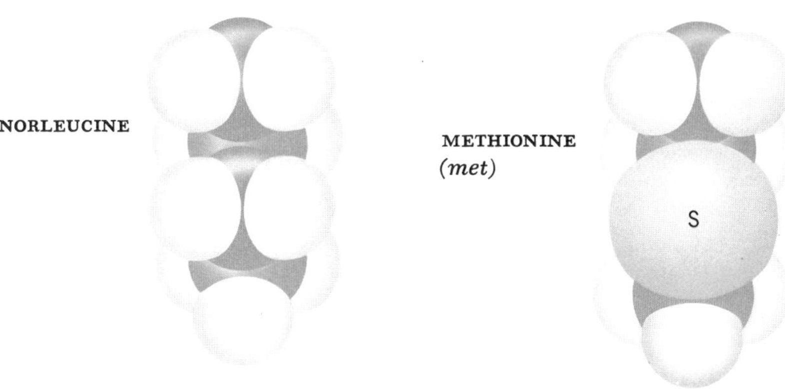

Norleucine as a hydrophobic side chain is absent and its place is taken by methionine. Where shape and hydrophobicity are critical, methionine can simulate norleucine, and its sulfur atom gives it additional ligand-forming properties.

right). Pro, the only amino acid in which the side chain loops back to reattach to the main chain, has the property of forcing a bend in the main chain and of disrupting an α helix. In myoglobin and hemoglobin, which are essentially made up of lengths of α helix connected by bends of extended chain, every bend does not have a Pro, but every Pro produces a bend. (See page 51 for an illustration of Pro as a helix breaker in myoglobin.)

Other residues have been examined for their influence on α helix formation, particularly in earlier days when the α helix was taken as the model unit of protein structure. It has been suggested that side chains which branch at the β carbon* would be so bulky as to make an α helix unstable if they occurred adjacent to one another on successive turns of the helix, or every third or fourth residue along the chain. By this criterion, Thr, Val, and Ilu would be helix breakers, along with Pro, Cys for its cross-linking proclivities, and perhaps Ser, Asn, and Gln because of their hydrogen-bond cross-linking. It is striking that most residues, no matter how complex their side chain, do have a compact $-CH_2-$ group at their β carbon, as though other side chains with bulk at the β carbon were weeded out long ago as too difficult to accommodate in proteins. These principles have been used to try to predict total α helix content from amino acid composition but, like other noncrystallographic methods of estimating helix content, have met with only limited success.

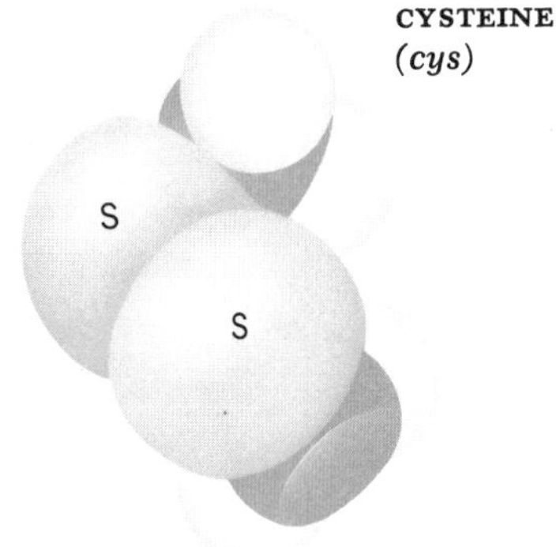

Cysteine provides both reactive —SH groups and polypeptide-chain cross-linking. It was once common, during protein degradations, to refer to two cysteines cross-linked as above as a cystine *residue.*

* The main chain atom to which the side chain is attached is called the α carbon, and successive carbons out along the chain are denoted β, γ, δ, and so on. The side-chain $-NH_2$ group of lysine (*right*), being attached to the ε carbon, is often called the ε amino group in contrast to the α amino group that participates in the peptide bond.

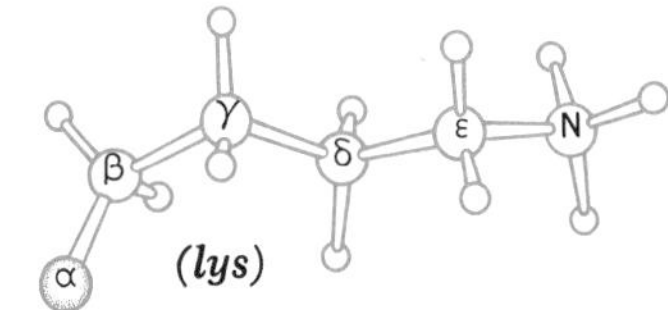

Adenine

Thymine (Uracil in RNA, without CH_3)

Guanine

Cytosine

The four bases of DNA, paired as above, are the four letters in the alphabet of the genetic code.

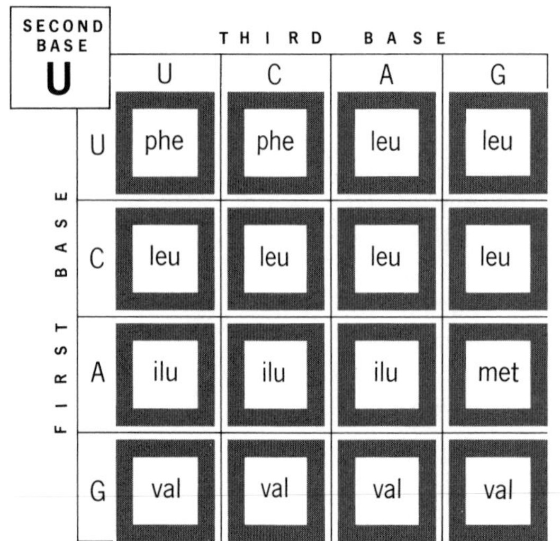

SECOND BASE **U**

FIRST BASE	THIRD BASE U	C	A	G
U	phe	phe	leu	leu
C	leu	leu	leu	leu
A	ilu	ilu	ilu	met
G	val	val	val	val

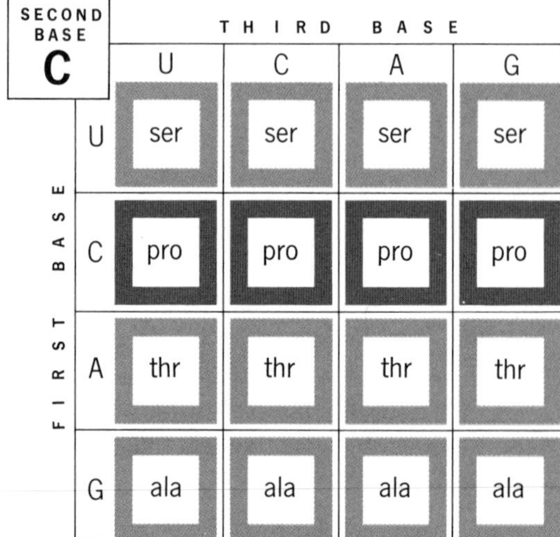

SECOND BASE **C**

FIRST BASE	THIRD BASE U	C	A	G
U	ser	ser	ser	ser
C	pro	pro	pro	pro
A	thr	thr	thr	thr
G	ala	ala	ala	ala

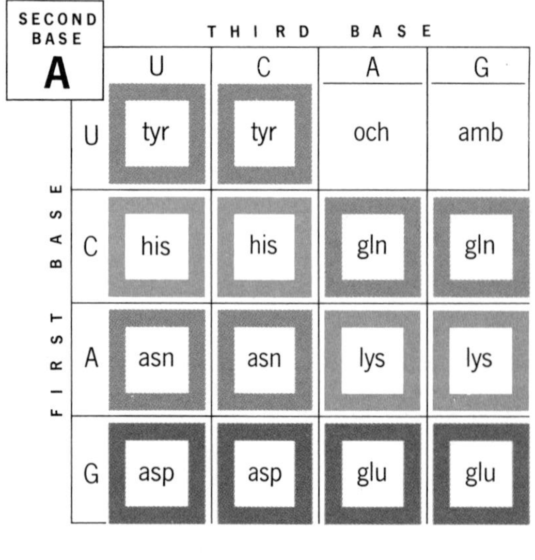

SECOND BASE **A**

FIRST BASE	THIRD BASE U	C	A	G
U	tyr	tyr	och	amb
C	his	his	gln	gln
A	asn	asn	lys	lys
G	asp	asp	glu	glu

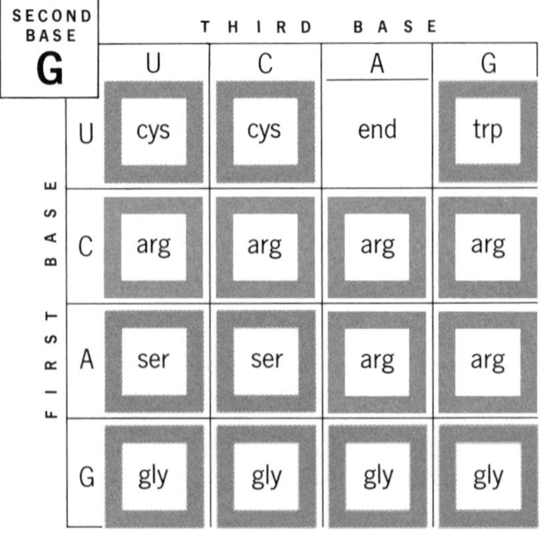

SECOND BASE **G**

FIRST BASE	THIRD BASE U	C	A	G
U	cys	cys	end	trp
C	arg	arg	arg	arg
A	ser	ser	arg	arg
G	gly	gly	gly	gly

Hydrophobic — Ambivalent — Basic hydrophilic — Acidic hydrophilic

One of the recent triumphs of molecular biology has been the laborious working out of the nucleic acid code—the pattern of three successive bases on the molecule of DNA or RNA, which tells which amino acid is to come next in the synthesis of the polypeptide chain (19). With the recent identification of a third "stop" code (20), the table is apparently complete as shown on the opposite page. Each of the 64 possible permutations of four alternate bases at three sites codes for an amino acid except UAA and UAG, the so-called "ochre" and "amber" mutants (named for Brown and Bernstein, respectively), and UGA, all of which apparently tell the ribosome to stop synthesizing the chain. Several regularities appear from an examination of the code. The first is that the purines, A and G, or the pyrimidines, U or C, are often interchangeable in the third base position along the nucleic acid, and that in many cases it makes no difference at all what the third base is. Thus CUU, CUC, CUA, and CUG all produce Leu. It also appears that the second of the three bases is most essential for determining the character of the residue concerned. U in the second position ensures that the residue will be nonpolar—an "inside" residue. With a C in the second position, the residue will be either nonpolar or a neutral polar group that can be accommodated in a hydrophobic environment. Only with A or G in the second position will a residue be produced which *must* be on the surface of the molecule. The basic groups have C or A in the first position, A or G in the second, and any of the four in the third. The other residues in this immediate vicinity in the table are the most polar of the uncharged groups, Asn, Gln, and Ser. The acidic residues are confined to GAX, where Ẋ is any base. The "stop" code is a uracil followed by any two purines except GG, and UGG itself codes for a highly unusual amino acid which can be obtained by no other triplet combination.

These regularities in the amino acid code, of course, are themselves the result of natural selection very far back in the history of living organisms. The code is to a certain extent a "fail-safe" system. If the table on the opposite page is visualized as a 4 by 4 by 4 cubic matrix, then one single-base mutation will be represented by a shift along a row, column, or layer parallel to one of the three principal axes. If a given residue is initially Phe, Leu, Ilu, Met, or Val, *any* mutation in the first or third base will produce another hydrophobic residue of hopefully not too disruptive an effect on the protein. If one of the codes for Arg mutates in the third base, it has a good chance of still producing Arg, and mutations in the first and second base also have a fair probability of producing Arg, Lys, or His. Such a code as actually exists could be called a conservative or self-correcting code. It would confer a considerable survival advantage upon its bearer in competition with some other life form with a more randomized code, for which most mutations would produce an amino acid with quite different properties.

CHAPTER TWO
BRICKS AND MORTAR—THE STRUCTURAL PROTEINS

The structural proteins have to be simple and repetitive for the same reason that all bricks in a building are alike—the instructions for building the edifice would be unmanageably complex otherwise. The raw materials are simple: a polypeptide chain and a selection of side groups. Only the succession of side groups is coded in the nucleic acids. It is the kind and grouping of side chains which make one chain tend to fold into an α helix of hair keratin, another into β sheets of silk, and yet another into the collagen triple helix.

Most of the folding schemes found in globular proteins have been encountered in more perfect form in one or another of the structural proteins. As we look briefly at these proteins, three questions should be kept in mind:

1. Why is a given folding plan stable?
2. How do the observed properties arise from the folding?
3. How does the sequence of amino acids produce the folding?

None of these questions can be given an unequivocal answer yet, and all are the subjects of active research. Each new protein structure makes the possible answers clearer.

2.1 THE LIMITATIONS ON FOLDING

If the peptide bond is planar, then the polypeptide chain has only two degrees of freedom per residue: the twist about the C_α—N bond axis, φ, and that about the C_α—C axis, ψ, as defined on the opposite page. (Standard angle definitions of reference [3] differ from older ones of references [1] and [2].)* A list of all the (φ,ψ) values for all residues will completely define the chain path. In fibrous proteins, the same (φ,ψ) values recur over long stretches of chain.

* Apparently yet another shift in definitions is being considered by the IUPAC–IUB Commission on Biochemical Nomenclature. This will achieve consistency with the usage of organic chemists, but will make the literature virtually unreadable. The new (φ,ψ) values can be obtained from the ones in this book by subtracting 180° from both angles. According to traditional sources, there are two types of sin: sins of omission and sins of commission. This is most definitely a sin of Commission.

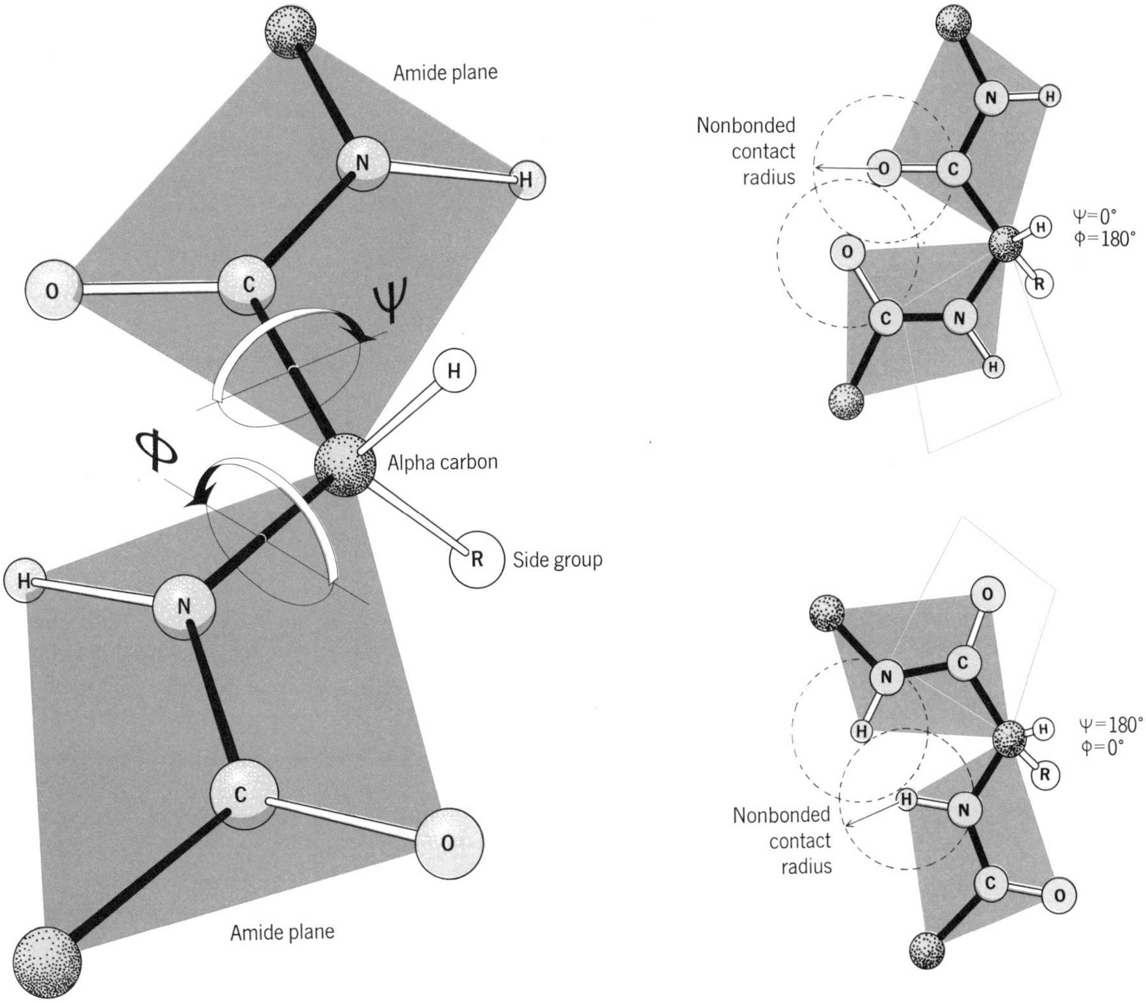

Amide plane

Ψ

Φ

Alpha carbon

Side group

Amide plane

Nonbonded contact radius

$\Psi = 0°$
$\Phi = 180°$

$\Psi = 180°$
$\Phi = 0°$

Nonbonded contact radius

Two amide planes are joined by the tetrahedral bonds of the α carbon. The only easy movement is rotation about the C_α—C and C_α—N single bonds. The rotation parameters are φ (C_α—N bond) and ψ (C_α—C). Positive rotation for φ and ψ is clockwise when viewed from the α carbon. The zero position in both φ and ψ occurs with the two peptide planes themselves coplanar as shown in the large drawing (above left). Some positions of φ and ψ are not permitted because of too close approach of unbonded atoms. The drawing at the top right shows the maximum forbidden overlap of carbonyl oxygens at φ = 180°, ψ = 0°. Below it is shown the maximum forbidden overlap of hydrogens at φ = 0°, ψ = 180°.

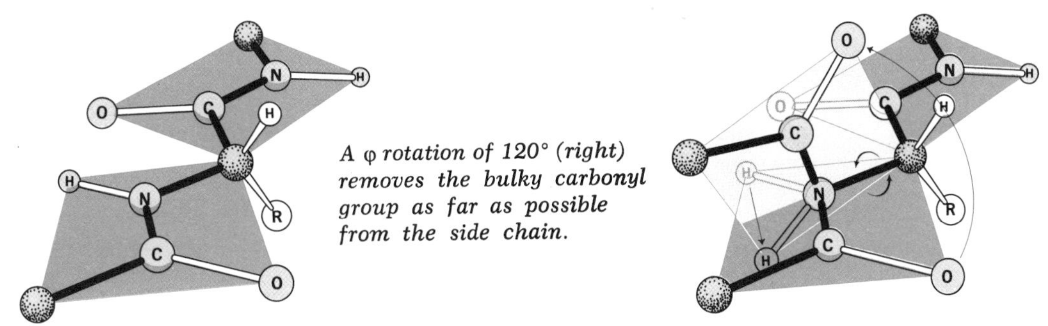

A φ rotation of 120° (right) removes the bulky carbonyl group as far as possible from the side chain.

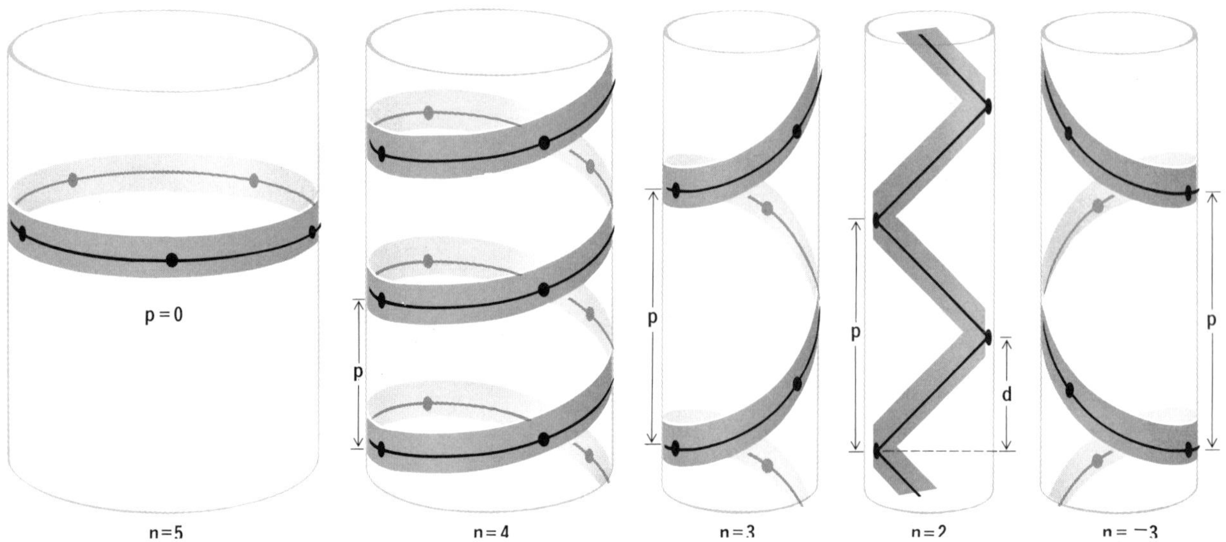

Definitions of the pitch of a helix, p, *and the number of repeating units per turn,* n. *The rise along the helix axis per repeating unit is* d = p/n.

Minimum contact distances for non-bonded atoms. Top figure is from normally accepted van der Waals radii. Figures in parentheses are the absolute minimum values found by Ramachandran in small structures. Distances in Å.

	C	N	O	H
C	3.20 (3.00)	2.90 (2.80)	2.80 (2.70)	2.40 (2.20)
N		2.70 (2.60)	2.70 (2.60)	2.40 (2.20)
O			2.70 (2.60)	2.40 (2.20)
H				2.00 (1.90)

If the twists at every α carbon atom are the same, then the chain falls naturally into a helix. Such a helix of repeating subunits can be described by the number of units per turn of helix, *n,* and by the distance traversed parallel to the helix axis per unit, *d* (above). The product of these is the pitch of the helix, *p*. For a polypeptide chain of fixed dimensions, both *n* and *d* are determined once φ and ψ are specified. The plot on the opposite page shows how the number of residues per turn varies with φ and ψ. Some foldings will be easily achieved; others will be impossible because they bring neighboring unbonded atoms too close together. G. N. Ramachandran and his group at Madras have studied possible conformations carefully, using models and computers. With the normally accepted atomic radii of the table at bottom left (1), only those conformations are permissible which fall within the two white regions of the so-called Ramachandran plot on the opposite page. On the other hand, with only a small decrease in minimum contact distances, these two allowed regions expand as shown by the lighter color tint, and a new allowed region appears near (240°, 240°).

The region from φ = 20° to 140° is particularly favorable. As the drawings on page 25 show, this corresponds to a rotation of the N—H bond toward the side chain and the removal of the CO group of the same peptide as far as possible from the side chain. As an example of the physical meaning of the Ramachandran plot, let us fix φ near 120° and see what happens as ψ varies through permissible and disallowed zones.* On the way, we shall encounter most of the folding schemes to be found in proteins.

*** The same hydrogen-bonded helix can be built with greater or lesser ease for a range of (φ, ψ) values along a line of constant *n*. The alpha helix, for instance, has been reported with (φ, ψ) values ranging from (132°, 123°) (1) to (113°, 136°) (4). The value φ = 120° falls within the range of variation of the helices discussed here.**

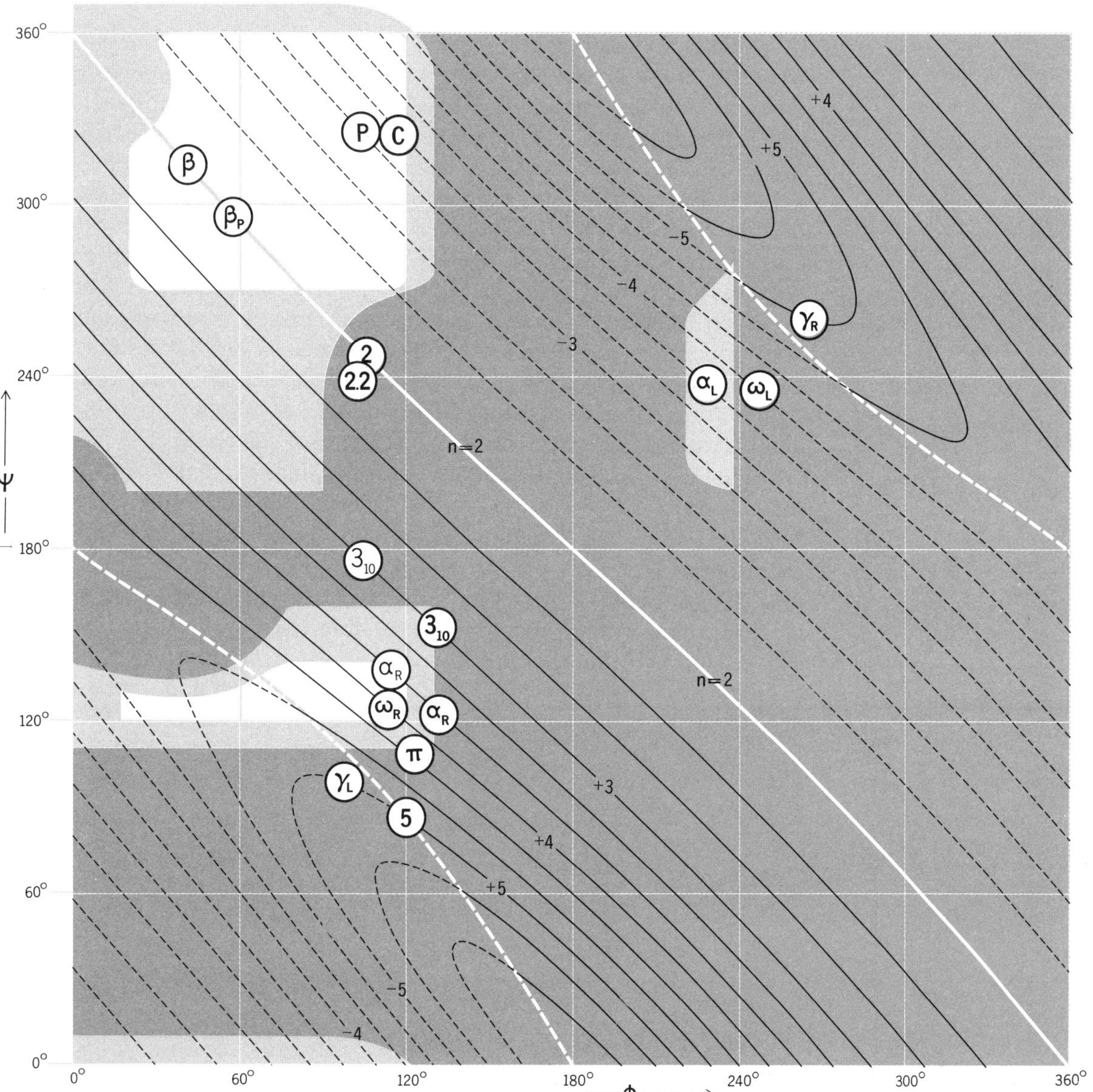

RAMACHANDRAN (φ, ψ) PLOT WITH CONTOURS OF EQUAL n

Right-handed helices

Left-handed helices

Twofold ribbon structures (White line is boundary between right- and left-handed helices.)

Ring structures (White line is boundary between right- and left-handed helices.)

Symbols in circles mark points on plot of proposed chain conformations. Alternate α helix (light type) from reference (4), alternate 3_{10} helix from reference (2). Others from reference (1).

- α_R *Right-handed α helix*
- α_L *Left-handed α helix*
- π *π helix (4.4 residues per turn)*
- 3_{10} *3_{10} threefold helix*
- 5 *Five-membered ring*
- γ_R γ_L *Right- and left-handed γ helices*
- ω_R ω_L *Right- and left-handed ω helices*
- β *Antiparallel β pleated sheet (silk)*
- β_P *Parallel-chain β pleated sheet*
- 2.2 *2.2_7 helix*
- 2 *2_7 ribbon*
- P *Polyproline helix*
- C *Collagen coiled coil*

The chart below is a plot of n and d against ψ for a value of $\varphi = 120°$. At $\psi = 86°$ the chain forms a flat ring with five residues. In this structure the carbonyl oxygens are brought into touching contact as they fold in toward the center on one side, and the structure is marginally unstable. As ψ increases, the ring opens and becomes a right-handed helix, and the number of residues per turn, n, decreases. Certain of these structures have particularly favorable hydrogen bonding from one turn of the helix to the next, and have been proposed and named in the past as models for protein structure. The hydrogen bonding of some of these is shown on the next page, and their structures on the next few pages. The π helix has 4.4 residues per turn, and the very important α helix has 3.6. The helix coils tighter around its axis as ψ increases, with five, then four, three, and finally two residues per turn. At the same time, the chain stretches out along its axis as d rises from zero for the rings through 1.5 Å for the α helix to 2.6 Å for the twofold ribbon. As this happens, the carbonyl groups, which had initially tilted inward toward the axis, rotate out.

At 3.6 residues per turn a particularly stable structure is reached, the α helix. The hydrogen bonding has dropped back one NH group from that of the π helix (opposite page, bottom). The carbonyl groups now lie parallel to the helix axis and point almost directly at the NH groups to which they are bonded.

The equations for n *and* d, *adapted from reference (2), are:*

$$\cos\left(\frac{\theta}{2}\right) = +\,0.817\sin\left(\frac{\varphi+\psi}{2}\right) + 0.045\sin\left(\frac{\varphi-\psi}{2}\right)$$

$$d\sin\left(\frac{\theta}{2}\right) = -\,2.967\cos\left(\frac{\varphi+\psi}{2}\right) - 0.664\cos\left(\frac{\varphi-\psi}{2}\right)$$

$$n = 360°/\theta$$

θ is the rotation about the helix axis per repeating unit.

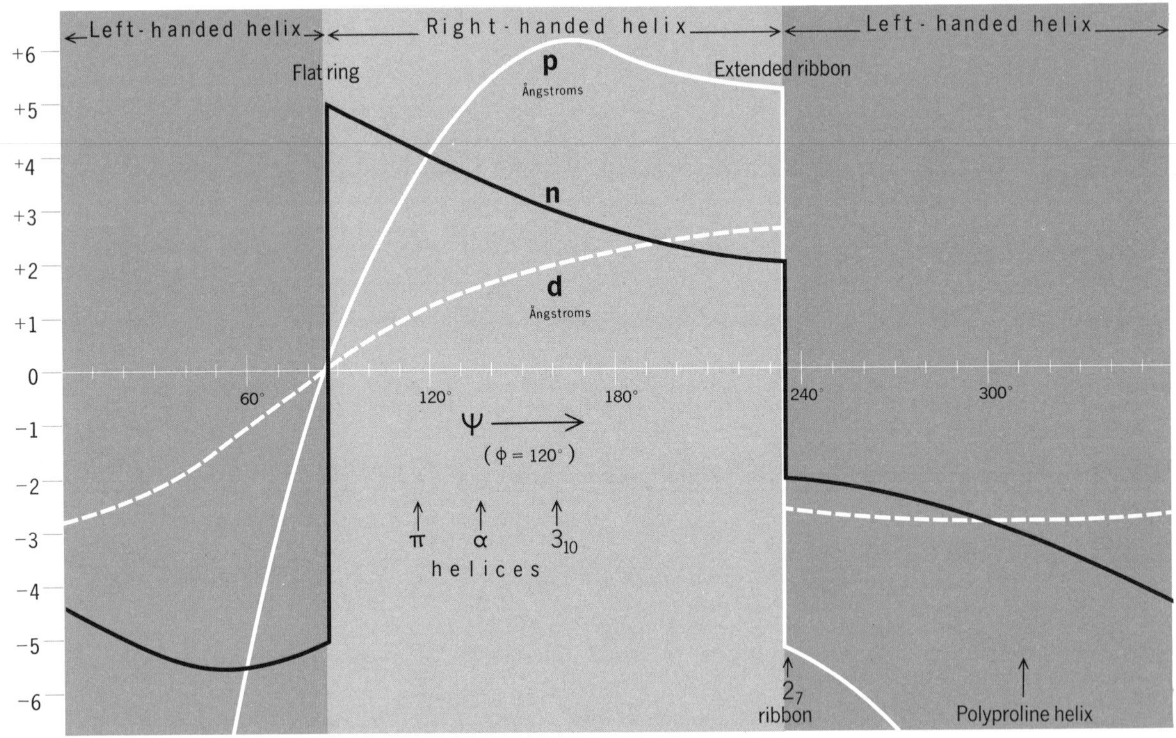

Plot of n, d, *and* p *versus* ψ *for* $\varphi = 120°$, *with several important helices marked.*

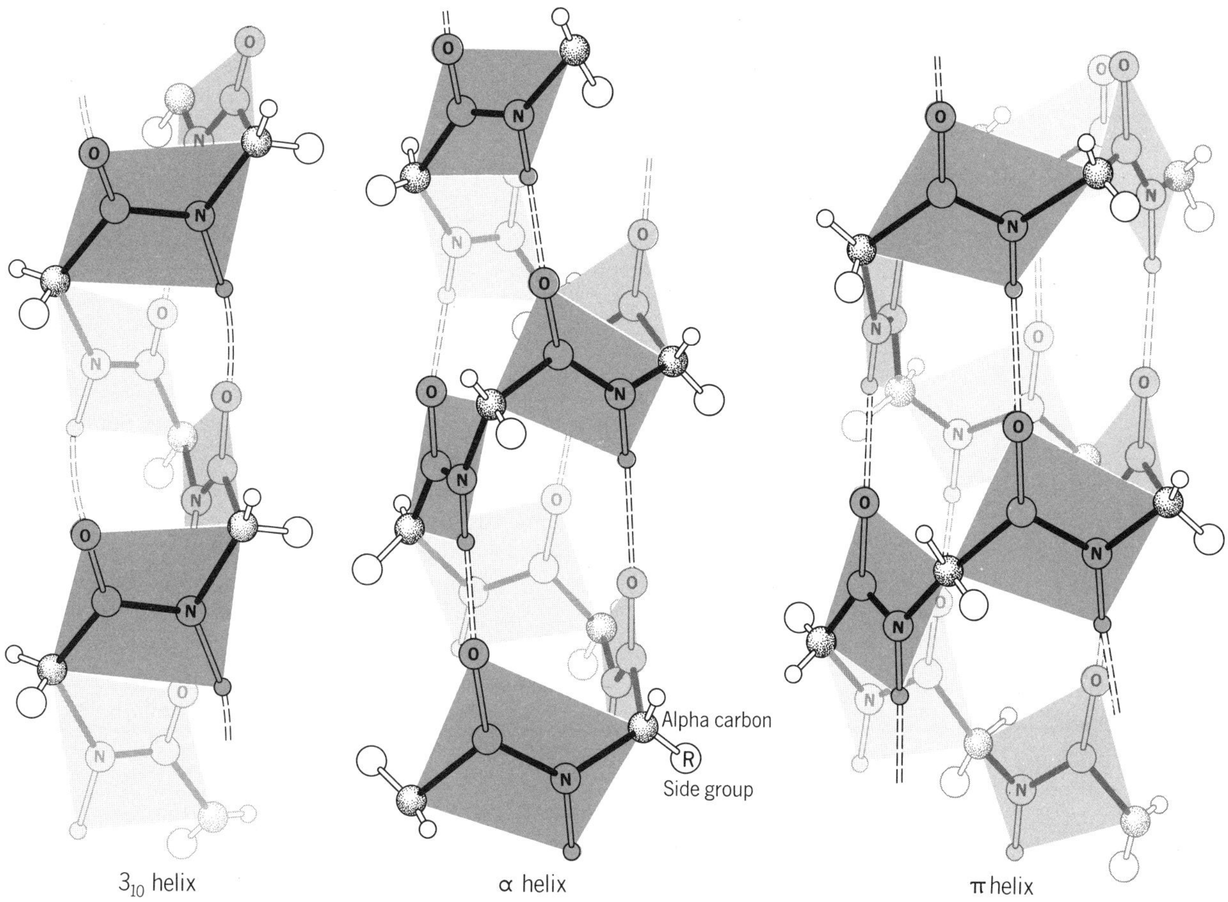

Alpha carbon
R
Side group

3_{10} helix   α helix   π helix

The 3_{10}, α, and π helices differ in their patterns of hydrogen bonding, as shown below. Hydrogen bonds in the α helix are particularly unstrained, making the α helix especially stable. The α carbons are stippled, with small attached spheres for hydrogens and larger spheres for side groups.

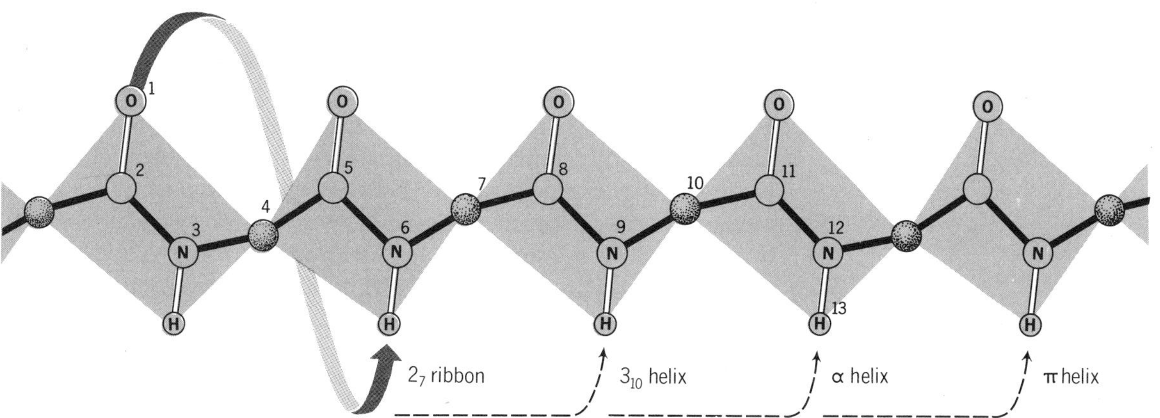

Hydrogen bonding pattern from one turn of the helix to the next, for four related helices. The principal number in the helix notation denotes the number of residues per turn and the subscript tells the number of atoms in the ring formed by closing the hydrogen bond. Thus the α helix is called the 3.6_{13} helix (see numbering above), and the π helix, the 4.4_{16} helix.

The physical bulk of the carbonyl oxygen and amide hydrogen prevent the configuration $\varphi = 120°$, $\psi = 210°$ from occurring.

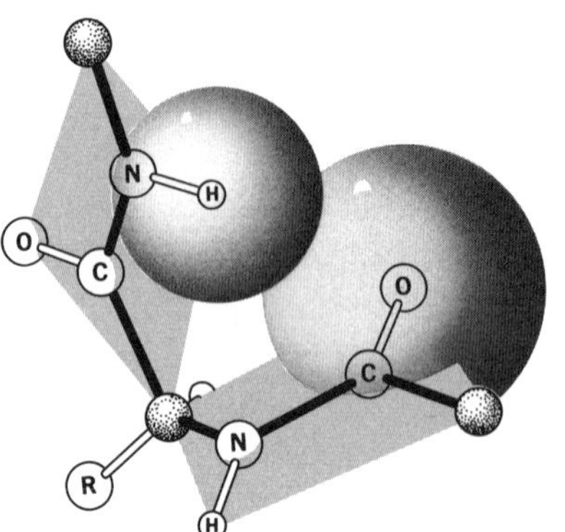

As ψ increases through 150° and beyond, another stable helix is reached, the 3_{10} helix, in which the carbonyl oxygen is bonded to a NH located one amide closer than in the α helix (page 29, bottom). The carbonyl groups point out from the helix axis, and the hydrogen bond is therefore bent and not as favorable. The α helix has been found in a great number of proteins, and short lengths of 3_{10} helix have recently been found in hemoglobin and lysozyme (see Chapters 3 and 4).

Beyond $\psi = 180°$, trouble develops, as the carbonyl oxygens begin colliding with amide hydrogens one residue up the chain. At $\psi = 234°$ the pitch of the helix becomes infinite and the helix flattens out into a ribbon with alternation of residues to the right and left. This 2_7 ribbon (left), which has been found so far only in a slightly modified form in one tetrapeptide and never in a protein, is tolerable only if the too-close oxygens and hydrogens are hydrogen bonded, and even then produces a shorter-than-normal hydrogen bond.

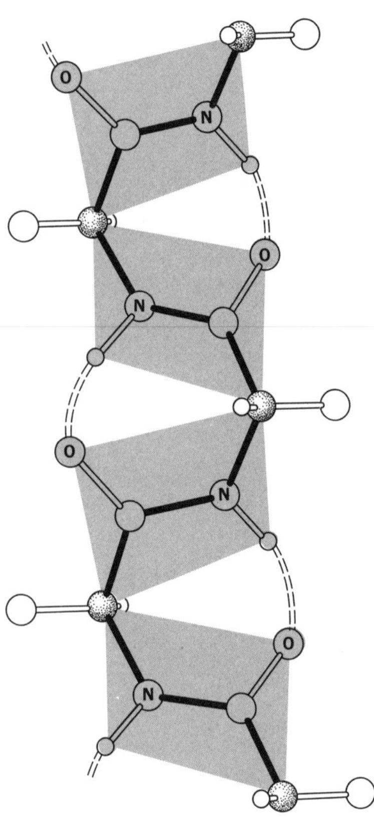

Part of the O · · · H strain in the 2_7 ribbon shown above can be relieved by warping the ribbon into a gentle 2.2_7 helix.

Above $\psi = 234°$ a new zone of behavior is entered, and the helix develops a left-handed sense. Near 310° it has three residues per turn, a structure that has been found in polyproline (right). There is one important difference in these new left-handed structures: the carbonyl and amide groups are oriented almost perpendicular to the helix axis and are in no position to form hydrogen bonds with groups on the same chain. Instead, in crystals of polyglycine and polyproline, hydrogen bonds from one chain to another in a direction perpendicular to the chains knit the crystal structure tightly together. In collagen, three such "polyproline" helices twist slowly around one another in a right-handed superhelix, with one or two out of every three carbonyl and amide groups involved in hydrogen bonding between chains. The effect that such bonding has on physical properties will be discussed later.

As ψ increases still more, the number of residues per turn increases and the helix winds down toward the flat pentapeptide ring again. At 350° there are four residues per turn, but the carbonyl groups begin to clash around the helix axis. The region above $\psi = 20°$ becomes completely impossible, and the strain is just beginning to be relieved at $\psi = 86°$ when the initial pentapeptide ring reoccurs.

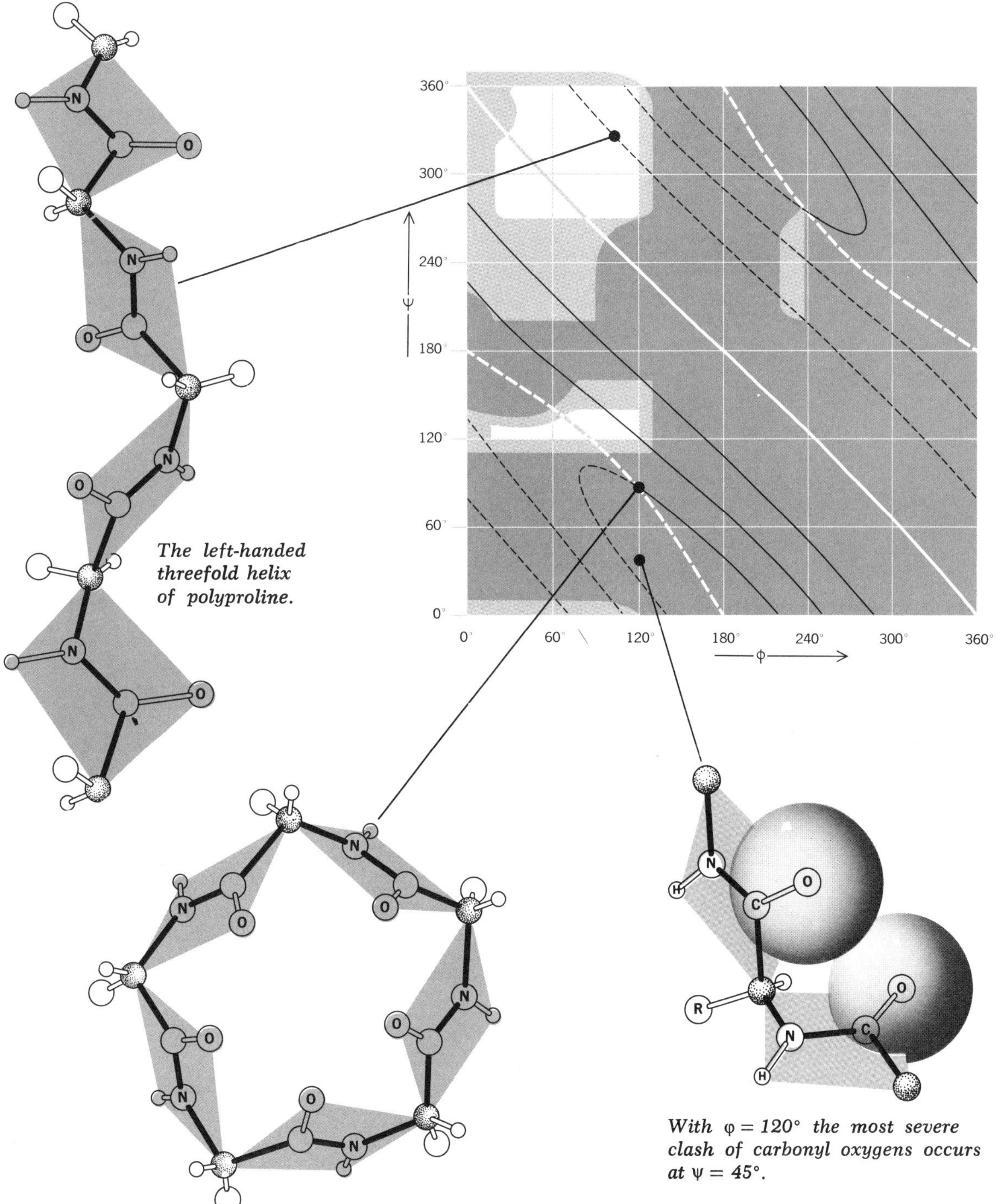

The left-handed threefold helix of polyproline.

With φ = 120° the most severe clash of carbonyl oxygens occurs at ψ = 45°.

The five-membered ring at φ = 120°, ψ = 86° is barely acceptable, as the carbonyl oxygens touch.

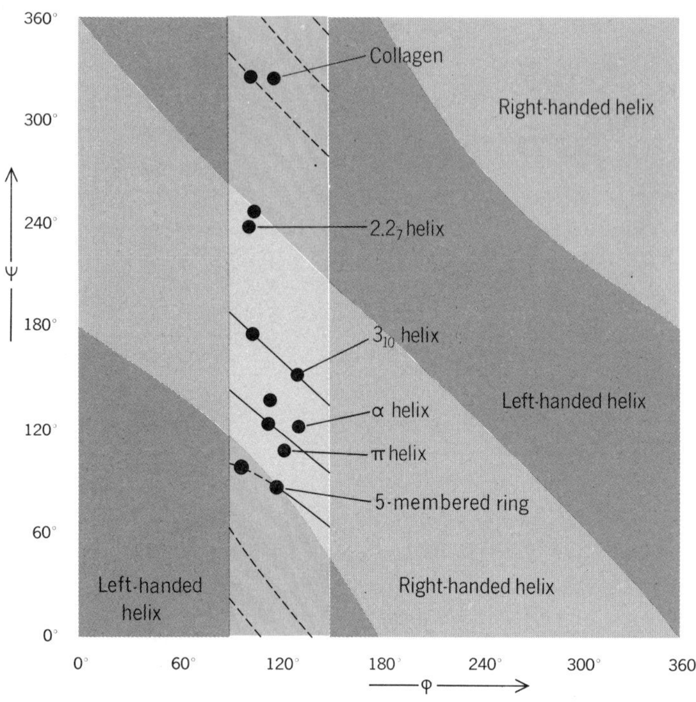

Thus, for $\varphi = 120°$, there is a right-handed internally bonded helix region from $\psi = 86°$ to 234°, a mildly forbidden buffer zone around the 2_7 ribbon, a left-handed externally bonded region from 234° through zero to 20°, and a totally prohibited zone from 20° to 86°. Virtually all of the configurations found in proteins fall in these regions. As φ falls below 120°, the boundaries of these zones change in the way shown above, but their basic character remains the same. If the side chains were of no more bulk than hydrogen atoms, then the Ramachandran plot would be symmetrical through its center; a similar statement could then be made for the right half of the plot, with left-handed α helices and right-handed externally bonded helices. But this is not so except for Gly, and for other amino

Structural parameters of five important types of protein-chain conformations. Alternate parameters for α helix from (4), and for 3_{10} helix from (2). All others from (1).

	α helix	3_{10} helix	2_7 ribbon	Polyproline helix	Antiparallel β pleated sheet
φ	132°(113°)	131°(106°)	105°	103°	40°
ψ	123°(136°)	154° (176°)	250°	326°	315°
n	3.61	3.00	2.00	−3.00	2.00
d (Ångstroms)	1.50	2.00	2.80	3.12	3.47
p (Ångstroms)	5.41	6.00	5.60	9.36	6.95

acids, collisions with the side chain keep almost all of the right half of the plot forbidden except for the vicinity of the left-handed α helix.

The actual distribution of configurations about the 129 α carbons of the enzyme lysozyme is shown on page 74. At the moment, note only the clustering of points in the left half of the plot between $\psi = 100°$ and 220°, and again between 250° and 360°, the narrowness of the forbidden zone around the 2_7 ribbon, and the wide forbidden zone from $\psi = 10°$ to 100°. The crosses in the right half of the plot are glycines, which have no side chain and are therefore exempt from the restrictions of the previous paragraph. Only a few scattered residues have φ values above 150°, and most of these are clustered near the left-handed α helix.

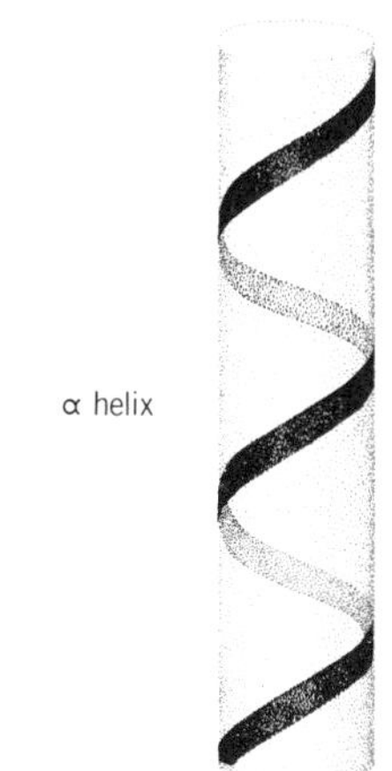
α helix

Now let us see which of these structures are used in the different structural proteins, and why.

2.2 THE VARIETIES OF FIBROUS PROTEINS

The most familiar of the fibrous proteins are probably the keratins, which form the protective covering of all land vertebrates: skin, fur, hair, wool, claws, nails, hooves, horns, scales, beaks, and feathers. Equally widespread if less visible are the actin and myosin of muscle tissue. Epidermin is another skin component, and fibrinogen is the precursor of the blood-clotting mechanism. A second great class of proteins are the silks and insect fibers. A third class are the collagens of tendons and hides, which form connective ligaments within the body and give extra support to the skin where needed.

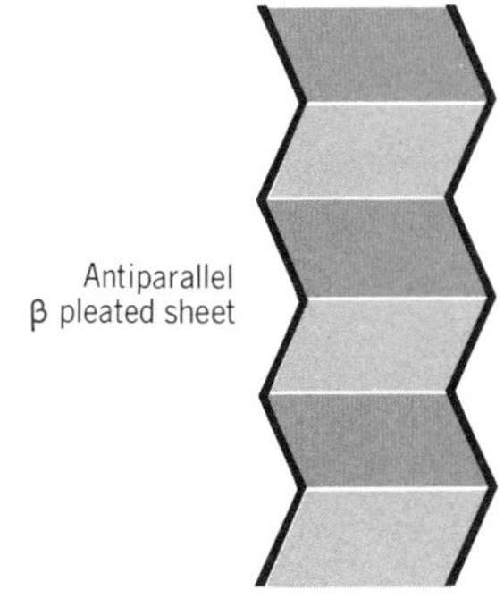
Antiparallel β pleated sheet

These proteins are built up from three main structures: the α helix, the antiparallel β pleated sheet, and the triple helix. The keratins are mostly α helix, with feathers in birds and some of the stiffer parts in nonmammals being a complicated form of β sheet. Myosin, epidermin, and fibrinogen are also α helical. The silks are the best example of the β sheet, and collagens use a characteristic triple helix.

Wool fibers form perhaps the best example of an α helical structure. They are flexible, and are extensible over a long range, up to twice their normal length. Yet they are elastic; when the tension is released the fibers snap back again. It is just these properties that are responsible for the springiness and live feel of good wool cloth. On the other hand, the fibers are only moderately strong. The silk β sheet structure is characterized by unusual strength, great flexibility, but low extensibility. If tension is applied to a silk fiber, rather than stretching, it resists elongation up to a point and then breaks. The collagen triple helix is quite strong, resistant to stretching, and also relatively rigid.

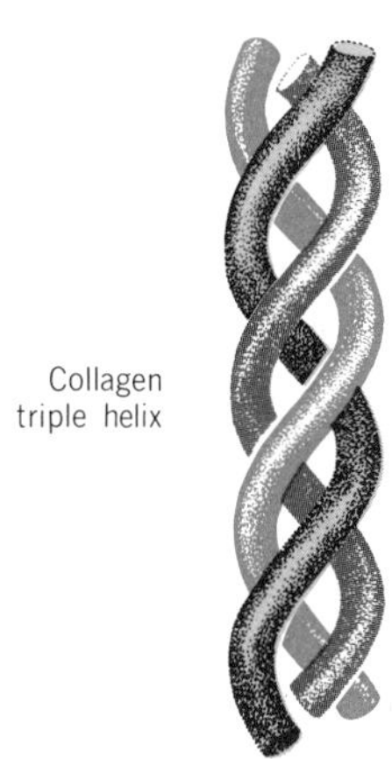
Collagen triple helix

Amino-acid content of three typical fibrous proteins, expressed in mole percent of amino acids. Principal components are marked with an asterisk. Silk fibroin is from the silkworm, Bombyx mori. *Wool keratin is from merino sheep and collagen from rat-tail tendon.*

	gly	ala	ser	glu + gln	cys	pro	arg	leu	thr
SILK	44.6*	29.4*	12.2*	1.0	....	.3	.5	.5	.9
WOOL	8.1	5.0	10.2	12.1	11.2	7.5	7.2	6.9	6.5
COLLAGEN	0*	10.7*	4.3	7.1	....	12.2*	5.0	2.4	2.0

The amino acid content of silk, wool, and collagen is shown above (data from reference [5]). Silk is primarily made up of Gly, Ala, and Ser in the ratio 3 : 2 : 1, with a small amount of other, mostly bulky side chains such as Tyr, Arg, Val, Asp, and Glu. In wool, the Gly, Ala, and Ser content is less than for silk, but there are many more bulky residues: Arg, Glu, Cys, Pro, Leu, Thr, Asp, and Val. Collagen is chiefly Gly, Ala, Pro, and Hypro with a few minor components. (The unusual amino acid hydroxyproline differs from proline in having a hydroxyl group on the side-chain ring. It is peculiar to the collagens and is not one of the 20 amino acids coded for in nucleic acids.)

How are these macroscopic properties explained by molecular structure?

2.3 SILKS AND THE β SHEET

Silks are built from extended polypeptide chains stretched parallel to the fiber axis, with neighboring chains running in opposite directions and hydrogen-bonded to form a sheet as shown below. Pauling and Corey, when they proposed this in 1951, called it the antiparallel pleated sheet (6). The need to make good hydrogen

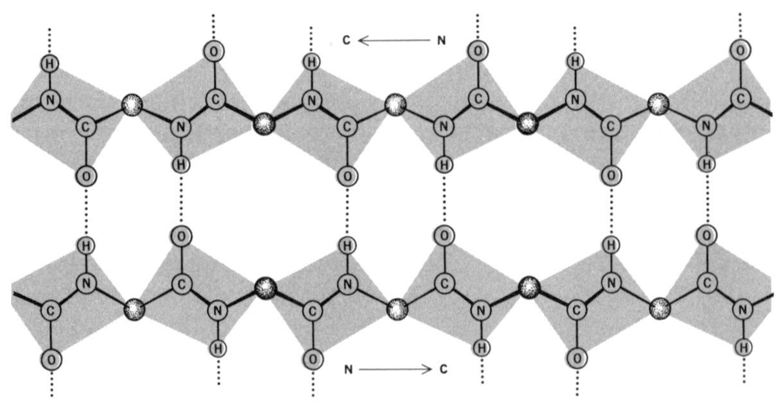

Hydrogen bonding within one antiparallel β sheet. Unterminated dotted lines are hydrogen bonds to neighboring strands in the sheet. Side chains are not shown.

asp + asn	val	hypro	hylys	tyr	ilu	phe	lys	trp	his	met
1.3	2.2	....	....	5.2	.7	.5	.3	.2	.2	....
6.0	5.1	....	....	4.2	2.8	2.5	2.3	1.2	.7	.5
4.5	2.3	9.4*	.7	.4	.9	1.2	2.7	....	.4	.8

bonds keeps the chain from being fully extended, and so the configuration falls at β on the Ramachandran plot (page 27) instead of at (0, 0). This produces the pleated effect visible in the perspective view at the bottom of the page, and pushes successive side chains far out away from the sheet on alternating sides.

Chemical sequence studies of digested silk have shown that a basic six-residue unit repeats for long distances in the chain (7):

$$(\text{Gly–Ser–Gly–Ala–Gly–Ala})_n.$$

This accounts for the peculiar amino acid composition of silk, and also means that all the Gly will be on one side of the β sheet and all the Ser and Ala on the other. The two-residue repeat distance down the polypeptide chain is 6.95 Å and the spacing between antiparallel chains in the sheet is 4.72 Å. These sheets then pack with glycines to glycines and alanines (and serines) to alanines, as on page 36, so the distance between sheets alternates between 3.5 Å and 5.7 Å (8).

The result is a fiber that is very strong because the resistance to tension is borne directly by the covalent bonds of the polypeptide chain. It is not appreciably extensible, for the chain is already stretched as far as it can go without breaking the hydrogen bonds that hold the sheet together. But since the sheets themselves are held together only by van der Waals' forces between unbonded side chains, silk is quite flexible.

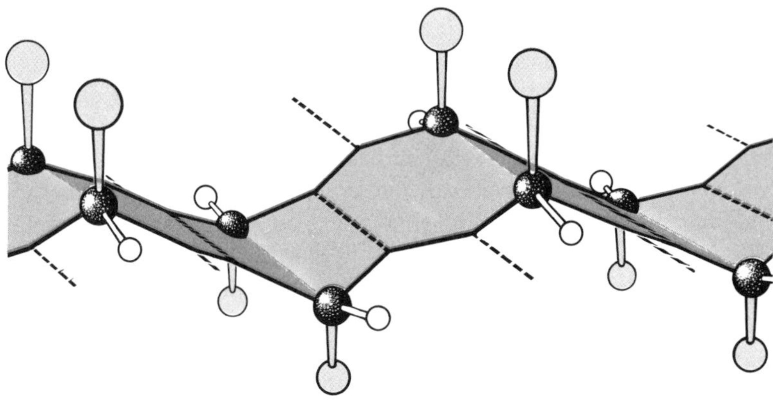

The β sheet is pleated or rippled, with alternate side chains extending out on either side. Dashed lines indicate hydrogen bonds.

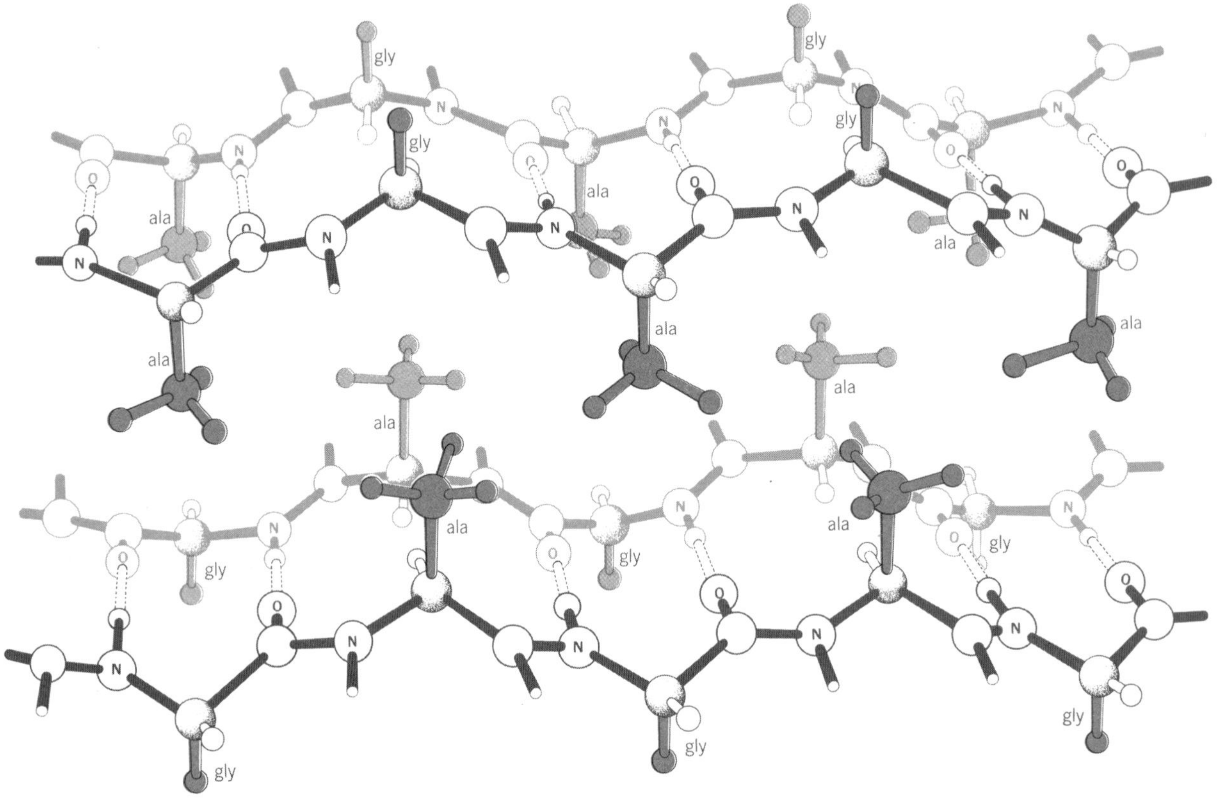

THE THREE-DIMENSIONAL ARCHITECTURE OF SILK. *The side chains of one sheet nestle quite efficiently between those of neighboring sheets. The cut bonds extend to neighboring chains in the same sheet.*

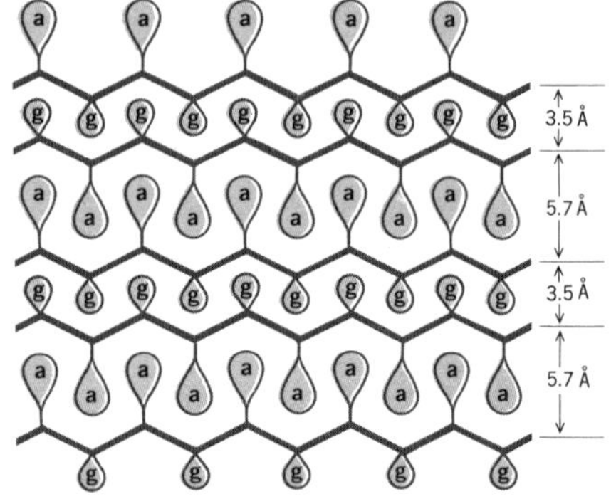

The sheets in silk are packed with ala against ala and gly against gly. The spacing between sheets therefore alternates between 5.7 Å and 3.5 Å.

But this is only part of the story. There is no room in this regularly packed crystalline structure for bulky side chains such as tyrosine. The ordered regions alternate with disordered regions which contain, in addition to the three primary residues, all the large side chains. These amorphous regions are responsible for what little extensibility silk fibers do have. In the most common silk from *Bombyx mori,* roughly 60 percent of the protein is crystalline. This does not imply separate polypeptide chains; one chain can run through several crystalline regions with disorder in between.

Different species of silkworm produce proteins with different proportions of bulky amino acids, and these have exactly the differences in physical properties that would be expected (9). *Anaphe moloneyi* silk has only 5 percent bulky groups and the most crystalline fibers of all. It remains elastic—its stretch remains proportional to the strain—up to its break point at the relatively low extension of 12.5 percent. The common *B. mori* silk has 13 percent bulky groups and about 40 percent noncrystalline regions. It remains elastic almost but not quite all the way to its break point at 24 percent extension. *Antherea mylitta* has the most amorphous material of all, with 29 percent bulky side chains. It begins to show a stress-strain curve more like that of wool keratin—initial elasticity, a yield point at an extension of about 5 percent, followed by flow and then by a setting or hardening, as the amorphous regions become extended and oriented, and finally a break point at around 35 percent extension.

Thus the physical properties of silk fiber can be explained by the bonding within its structure: covalent in the first dimension, hydrogen-bonded in the second, and weak van de Waals' attraction in the third, with an alternation of crystalline and disordered domains. It is the peculiar (Gly—Ser—Gly—Ala—Gly—Ala) sequence that produces the crystalline regions, and it is the proportion of bulkier groups that determines the ratio of ordered to disordered regions and the load-bearing properties of the fiber.

2.4 α KERATIN

It is not surprising that the α protein first studied thoroughly was the α keratin of such a salable commodity as wool (10, 11) rather than fibrinogen or even muscle myosin. We shall look at hair keratin as a type example of an α protein. The structure of myosin is presented in references (49) and (50), Chapter 5.

The basic unit of hair keratin is an α helix, in about the same sense that the basic unit of an Aubusson tapestry is a thread. X-ray pictures taken by W. T. Astbury in the early 1930s showed a distinctive scattering pattern which later proved to be that of an α helix. Yet clearly this was only the basic unit of a complex structure, and the unraveling of the details of this superstructure has required the cooperation of x-ray diffractionists, electron microscopists, and wool chemists for the last decade and a half.

The weakest link is the gap between the x-ray and electron microscope evidence. The smallest features that can be seen in the best micrographs are tiny "protofibrils," 20 Å in diameter, running the length of the hair. The x-ray diffraction pattern shows that α helices must be present, yet these small protofibrils are too large to be α helices. Details of the protofibril structure are still debated, but the best suggestion is that one protofibril is made up of three right-handed α helices arranged in a left-handed coil, with sufficient tilt that the repeat distance along the protofibril is shortened from 5.4 Å to 5.1 Å (right).

The rest of the story is apparent from electron and light microscopy. As can be seen to the right and in the micrograph on the next page, nine protofibrils are bundled in a circle around two more to form an eleven-stranded cable, the "microfibril." This "nine plus two" arrangement has been found in other fibrous proteins, such

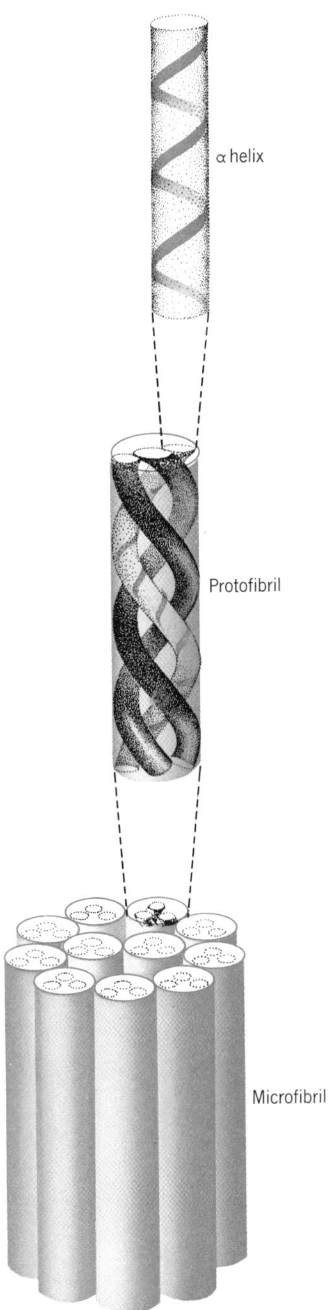

The assembly of hair keratin.

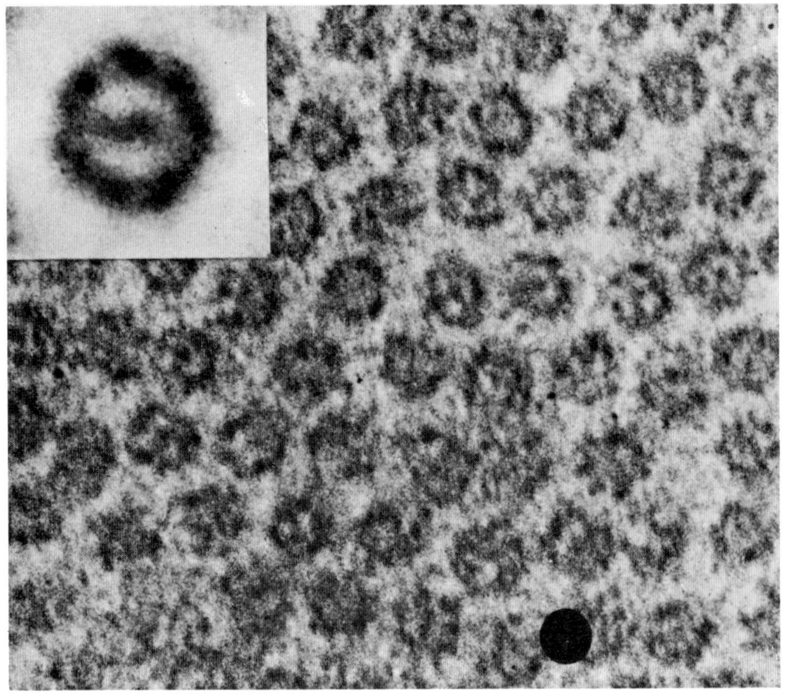

Electron micrograph of a cross-section of a merino wool fiber showing the microfibril structure. Inset: Superimposed print of several microfibrils in proper register to bring out details and to cancel background fluctuations. The black sphere in the main photograph shows the size of a hemoglobin molecule for comparison (By courtesy of Dr. G. E. Rogers [14]).

as bacterial flagella and sperm tails. These microfibrils, 80 Å across, are embedded in an amorphous protein matrix of high sulfur content, and hundreds of such microfibrils are cemented into an irregular fibrous bundle called a macrofibril. A section through several adjacent macrofibrils is shown in the electron micrograph on the opposite page. A typical wool fiber, 20 microns or 200,000 Å in diameter, is made up of packed dead cells of the order of 20,000 Å across. Within these cells lies a residue of macrofibrils oriented parallel to the fiber axis and each about 2000 Å in diameter. Thus a hair presents an ordered progression of structures: hair fiber, cell, macrofibril, microfibril, protofibril, and α helix. Part of its properties comes from the α helix, but much more comes from the way in which it is used.

The α keratins are very extensible; a wool fiber can be drawn out to twice its original length. When this happens, the α helices stretch with a breaking of hydrogen bonds between turns of the helix, and an extended β chain structure is formed. The hydrogen bonds themselves would not be enough to return the fiber to its original state after tension was removed. But the helices are cross-linked with disulfide bonds from Cys residues, and these cross-links form both a resistance to stretch to begin with and a restoring force when the stress is removed. The α keratins are classed as "soft" or "hard" by their sulfur content: the low-sulfur keratins of skin and callus are flexible and much more extensible than the high-sulfur hard keratins of horns, claws, and hooves.

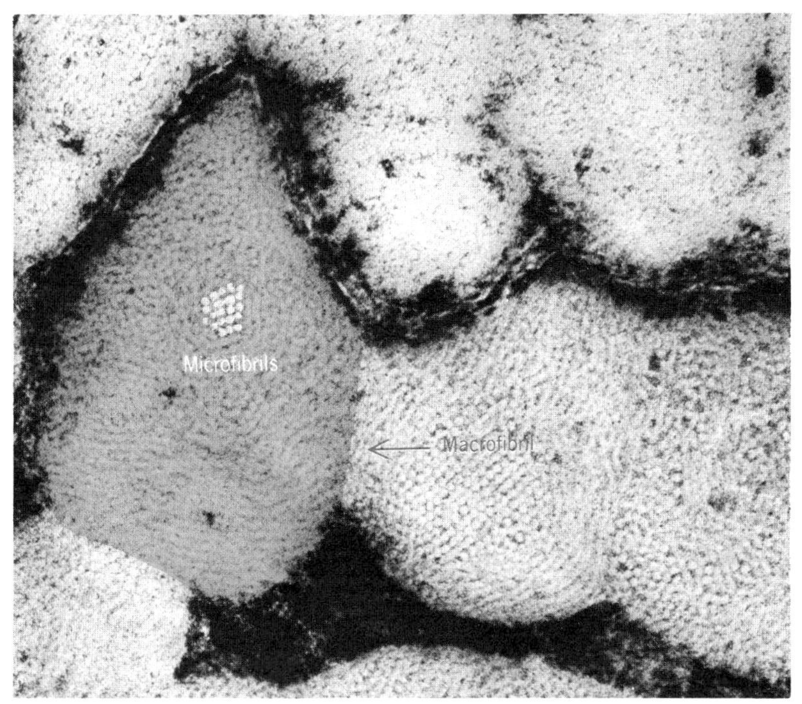

Wool keratin micrograph at lesser magnification, showing parts of several macrofibrils, one of which is colored, with microfibrils as light dots. The lighter microfibrils are embedded in a darker amorphous matrix which has absorbed more of the O_sO_4 stain [13].

Since the bonding in keratin is *within* one α helix, aside from the disulfide cross-linking, and the helices are packed in an amorphous matrix, α keratin is flexible. If it were not for these cross-links, the helices could break at occasional places under stress and then slide past one another upon stretching, with no tendency to return upon release. But the cross-links are frequent enough (the amino acid composition would suggest a maximum frequency of one cross-link every four turns of helix) that neighboring helices are locked together during the stretching process and relax back into their original arrangement afterward.*

The physical behavior of hair keratin, in summary, arises from the present of α helices, hydrogen-bonded within the chain, and cross-linked to a varying degree by disulfide bonds, embedded in an amorphous protein matrix. As any rope maker knows, coiled bundles and coiled-coil filaments and supercoils again are a sound way to make strong hawsers out of weak threads.

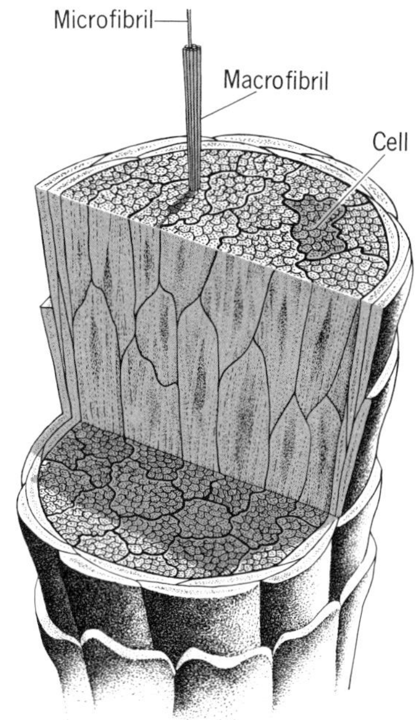

The organization of a complete hair.

---

* The basic principle of the permanent-wave process for hair is the breaking of the disulfide bonds between chains and the reforming of bonds after the hair fibers have been shaped as desired. A "permanent" is really permanent for the portion of the hair that was processed, and lasts until new and untreated keratin replaces it.

A hair follicle might not seem to be the most lively portion of anatomy, yet it is actually seething with activity. A reasonable growth rate of human hair of 6 in./year means that nine and one-half turns of α helix must be spun off every second.

Not all keratin is α keratin. Feather keratin shows an x-ray pattern that appears to be that of a rather complicated version of a β structure and one that has not yet been fully explained. There are also major β keratin components in the skin, scales, claws, and beaks of reptiles and birds, although the skin of amphibia is α keratin. β keratin may be an offshoot of α keratin which developed in the line of reptiles that led to birds but not in the Therapsid reptiles that produced the mammals.

2.5 COLLAGEN

The basic properties of collagen are rigidity and resistance to stretching. Collagen is used for connective ligaments where mechanical force must be transmitted without loss, in strengthening networks for the skin, and in other situations where a rigid, inert material is called for. As Borysko puts it (11, page 19):

> Collagen fibrils form a wide variety of patterns in the various tissues of the animal body. In cowhide, for example, they are grouped into fibers that branch and anastomose to form a complex structure ideally suited to resist the stresses and strains involved in holding together and protecting a ton of cow. . . . In the hind-leg tendon of the cow, the collagen fibrils are all oriented parallel to the long axis of the tendon. . . . In other tissues, such as the cornea, the collagen fibrils are arranged in laminated sheets, or they may occur as a fine network of individual fibrils holding the soft tissues and glands together. The mechanisms involved in the organization of the collagen fibrils in the tissues are completely unknown. . . .

A summary of what is known about collagen structure is shown on the opposite page. The fundamental structural unit of collagen appears to be a tropocollagen molecule, 15 Å in diameter and 2800 Å long, with a molecular weight of about 300,000. "Appears to be," for although the tropocollagen theory explains all the evidence at hand, no one has yet seen an isolated tropocollagen molecule in the electron microscope.

The tropocollagen molecules are thought to pack with a displacement of one quarter of their length to form an overlapping collagen fibril, having a characteristic banding pattern with electron-microscope stains every 640 Å. These fibrils are typically about 0.5 μ or 5000 Å in diameter but can be swollen to six or seven times this diameter in a weak acid medium for electron-microscope study.

The triple-helix structure of tropocollagen was worked out by Rich and Crick (15) and by Ramachandran and his group (16). The table on pages 34 and 35 shows that collagen is one-third Gly, with Pro and Hypro together making up another quarter. Polypro-

a

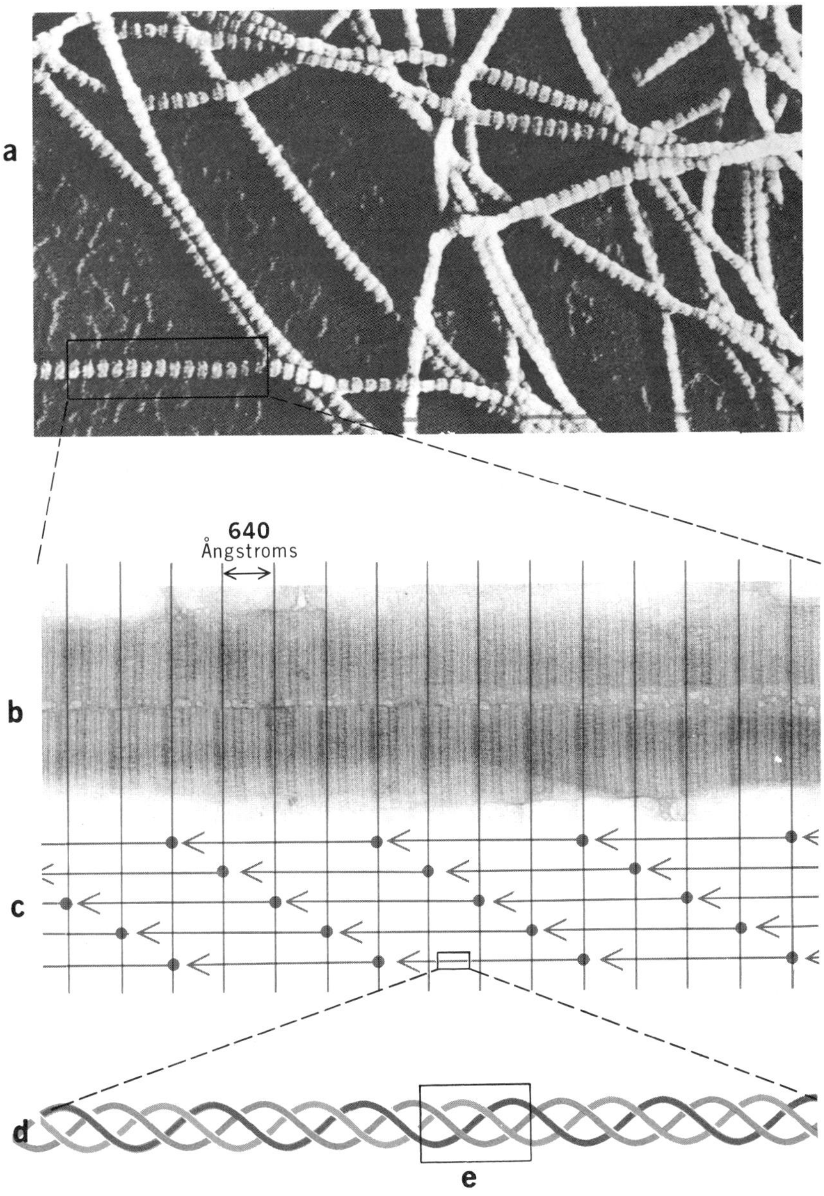

640
Ångstroms

b

c

d

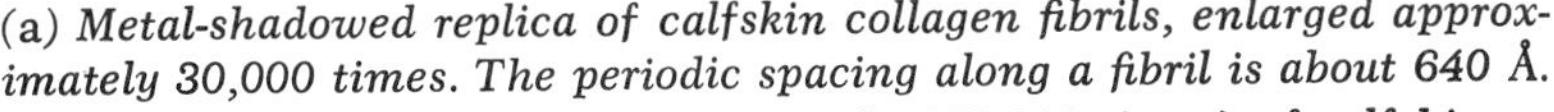

e

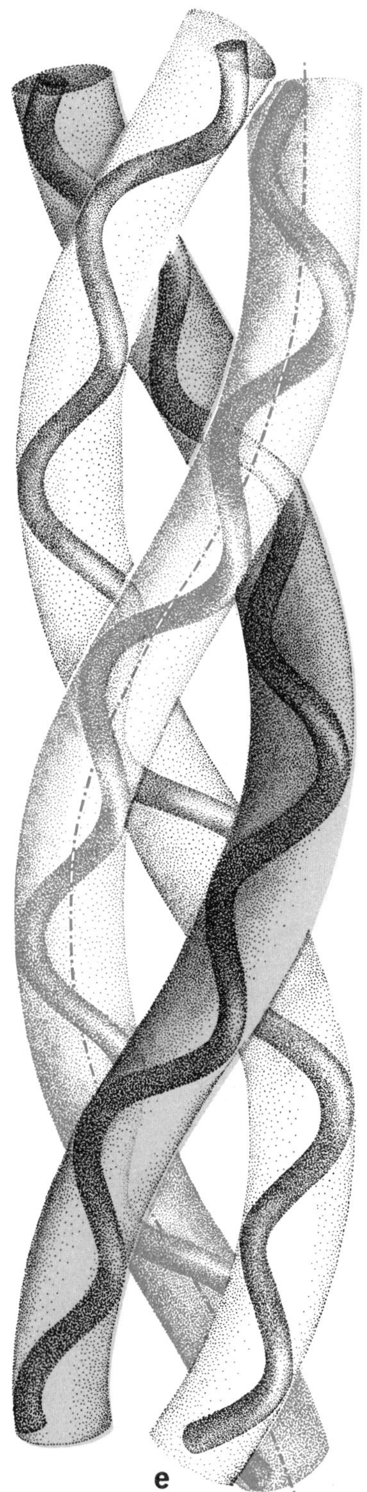

(a) Metal-shadowed replica of calfskin collagen fibrils, enlarged approximately 30,000 times. The periodic spacing along a fibril is about 640 Å.
(b) Higher magnification (approximately 100,000 times) of calfskin collagen fibrils stained with phosphotungstate. (Both photographs by courtesy of Dr. Alan Hodge.)
(c) One tropocollagen molecule, represented as an arrow, is believed to extend through four of the 640-Å long sets of cross-striations.
(d) The triple helix of the tropocollagen molecule.
(e) The section above, enlarged at the right, shows how three left-handed helices are given a right-handed twist to form a three-fold superhelix.

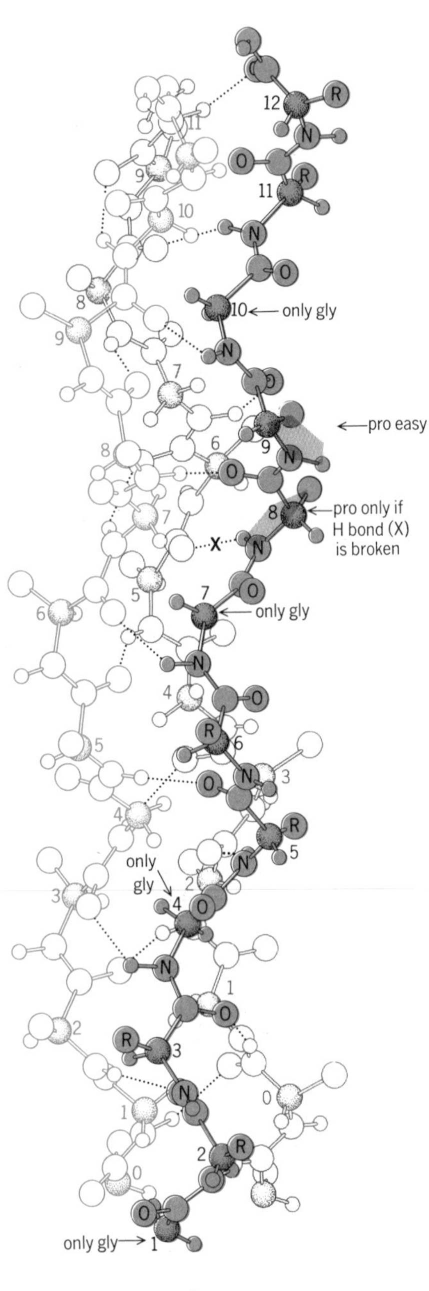

THE BASIC COILED-COIL STRUCTURE OF COLLAGEN

Three left-handed single-chain helices wrap around one another with a right-handed twist.

line, polyglycine, and polyhydroxyproline all give excellent fiber x-ray patterns indicating a three-residue extended helix (Ⓟ on page 27), with amides and carbonyls extending perpendicular to the polypeptide chain and hydrogen bonding from one chain to another. Three such parallel left-handed three-residue helices, removed from the crystalline environment and given a gentle twist to the right, form the basis of the collagen structure (left). Two of every three amides and carbonyls are hydrogen-bonded to one of the other two chains. All the α carbons are numbered. Those marked 1, 4, 7, . . . to the left are too close to the other two chains for side groups and can only be Gly, explaining the observed 33 percent Gly composition. The bulky pyrrolidine rings of Pro and Hypro can be accommodated at carbons 3, 6, 9, . . . without strain, and without eliminating an amide involved in hydrogen bonding. The sequences ⟮Gly—X—Pro⟯ or ⟮Gly–X–Hypro⟯, where X is any residue, are thus particularly favored. Little distortion is required to accommodate Pro at the intermediate carbons 2, 5, 8, . . . , although one hydrogen bond is lost for each Pro present.

The pattern of interchain hydrogen bonding makes the collagen triple-chain coiled coil strong and rigid, although the individual chains will collapse if the hydrogen bonding is broken. It is obvious that this particular structure is induced by amino acid sequences such as ⟮Gly—X—Pro⟯_n, ⟮Gly—X—Hypro⟯_n, and ⟮Gly—Pro—Hypro⟯_n, since polyglycine, polyproline, and polyhydroxyproline each prefer a folding very much like that of collagen. The fact that the tropocollagen molecules line up in the fibril with just the right quarter-length displacement indicates that there must be some long-range regularity in amino acid sequence beyond that which has been described here. Although many suggestions have been made for this, the matter is still unsettled. The amino acid sequence responsible for the triple helix is known; that which produces the higher level of organization is yet to be found.

2.6 SUMMARY

It has become apparent that there are three basic configurations of the polypeptide chain in fibrous proteins: the α helix, the collagen triple helix, and the β sheet. Structures of the most diverse function and properties are built up by adapting and varying these basic forms and by incorporating them into suitable kinds of higher organization. If a strong and rigid material is required, such as tendon, or the scales and swim bladders of fish, then the collagen helix is used, wound as it is into a triple-chain coiled coil with the chains tied tightly together by hydrogen bonds.

If a more flexible and elastic material is needed, the α helix is selected instead, with hydrogen bonds within each chain rather than from chain to chain. Extra elasticity and strength are achieved by winding the chains into coiled coils—apparently a triple strand followed by a "nine plus two" cable in keratins, where the emphasis is on strength, and a double strand in muscle myosin, where flexibility is required. Mechanical strength is enhanced even more in the α keratins by immersing the grouped cables in a matrix—a "laminated" construction. (One is reminded of the role of steel rods in pre-stressed concrete.) Finally, the most supple materials of all are the β sheets of silk, flexible yet strong, and resistant to stretching.

The arguments set forth above seem logical, yet there has certainly been a considerable amount of experimentation in the evolutionary process. As far back in the process as one cares to go (assuming, of course, that the parallel between lower ranks among the present phyla and early stages in the evolutionary process is legitimate), one finds the α helix present in muscle and collagen in tendons and fibers. On the other hand, the problem of protective surface layers has been solved in a different manner at different points. In earthworm cuticle a form of collagen is found, although without the banding ordinarily visible in electron micrographs. Fish scales are also members of the collagen class—elastoidin. Amphibian skin is α keratin. Reptilian scales and claws and avian beaks and feathers are something else again: feather or β keratin. And in the mammals, entirely analogous claws and nails, as well as skin, hair, and fur, are all α helical. It is tempting to see an arrival at a laminated α helical structure by a laborious trial and rejection of alternates.

The amino acid content and sequence determine which of the possible structures will be assumed. Pro and Hypro are incompatible with an α helix, and such a helix only occurs in the absence of large quantities of these residues. Polyglycine itself, lacking side groups, has been crystallized in both the β sheet and the polyproline helix, and the two classes with high glycine content are just the silks and collagen. Just as the sequence -(Gly—Ser—Gly—Ala—Gly—Ala)-_n induces a β sheet structure, so the sequences similar to -(Gly—X—Pro)-_n induce a collagen triple helix. It is less easy to say what sequence of amino acids will *produce* an α helix.* But in view of the great intrinsic stability of the α helix—its lack of steric strain and the perfection of its hydrogen bonds—it is probably fairer to list the factors that *prevent* it—sequences such as the above, large Pro concentrations, or large accumulations of bulky side chains that branch at the β carbon atom—and to regard the α helix as the most stable form in the absence of factors that specifically oppose it.

* X-ray analyses of synthetic polypeptides with repeating sequences analogous to those found in collagen (W. Traub, A. Yonath, and D. M. Segal, *Nature 221*, 915 [1969]) have provided strong support for the Rich/Crick model of collagen in which only one residue out of every three is hydrogen bonded to another chain.

CHAPTER THREE
MOLECULAR CARRIERS

3.1 THE OXYGEN-CARRYING TEAM

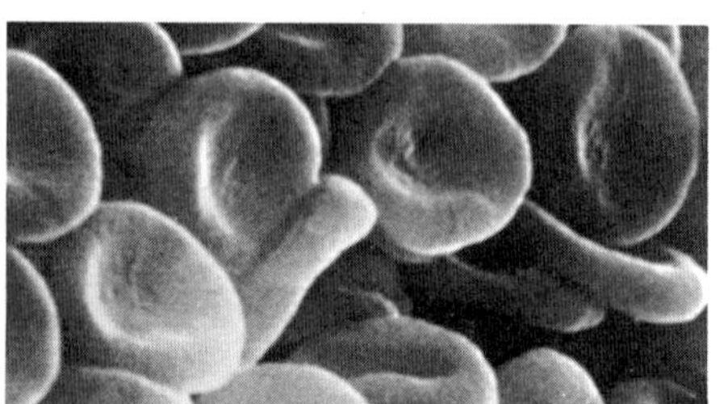

Scanning electron micrograph of human red-blood cells magnified approximately 3000 times. (By courtesy of Dr. Thomas Hayes.)

Two of the most important and plentiful proteins of vertebrates are hemoglobin and its cousin myoglobin. In the bloodstream there are something like 5 billion red cells or erythrocytes per milliliter (left). Each is packed with 280 million molecules of hemoglobin, of molecular weight 64,500 and with four heme groups per molecule.* In the tissues is myoglobin, the protein that is responsible for the deep red color of good steak. Myoglobin is much like hemoglobin, but is one quarter the size and has only one heme. The role of hemoglobin is to bind molecular oxygen to its heme irons at the lungs and to deliver it to myoglobin at the muscles; that of myoglobin is to store oxygen until it is required for metabolic oxidation. A second task of hemoglobin is to bring the carbon dioxide by-product of this oxidation back to the lungs and thus to prevent a buildup of acid in the muscles (1, 2).

The core of these molecules is the heme group: an iron atom surrounded by a porphyrin ring. The porphyrin ring is unsaturated in the sense of benzene and has many delocalized π electrons. Only one of many possible resonance structures is drawn to the left, and the backbone actually has fourfold symmetry, with bonds a and a', b and b', c and c', d and d' being equivalent.

The heme group is an essential component in hemoglobins, cytochromes, and enzymes such as catalase and peroxidase.

The hemoglobin/myoglobin pair has not always been the answer to the problem of distributing oxygen. One-celled organisms and organisms up to about 1 mm across can spread oxygen by simple molecular diffusion. Insects have surmounted the size barrier by developing hollow tracheal tubes, and can remain dependent upon diffusion. But this will work only up to a point. As J. B. S. Haldane has pointed out, this is one of the main reasons insects have remained small—and why the horror-film stories of giant man-eating ants are unlikely to come true.

* If an erythrocyte were enlarged 30 million times, it would be the size (and roughly the shape) of the Rose Bowl, piled high with 280 million large grapefruit.

The oxygen-carrying molecule problem has been solved again and again in the invertebrates, with such giant molecules as hemerythrin (mol. wt. 107,000; two nonheme iron atoms on each of eight polypeptide chains), the erythrocruorins (mol. wt. 400,000–6,700,000; one heme per 17,000), and chlorocruorin (mol. wt. 3,500,000; 190 chlorocruorohemes per molecule), all of which float free in the body fluid (3). But for vertebrates it seems to have been more efficient to build up a large package from smaller molecules, either because such a package functions better or because less genetic material is required to code for a small molecule. Myoglobin has one polypeptide chain of molecular weight 17,800 and one heme group. Hemoglobin has four similar chains of two types, two "α" and two "β", each with one heme.

Myoglobin is built from one heme group and a polypeptide chain of 153 amino acids.

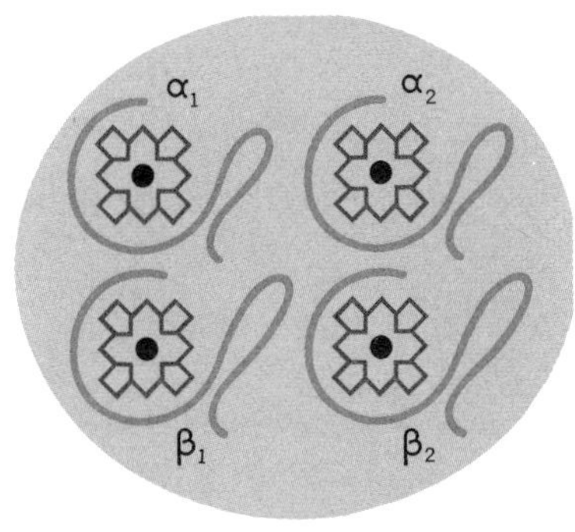

Hemoglobin has four hemes and four polypeptide chains, two α of 141 residues, and two β with 146.

The properties that are desirable in an oxygen-transporting system can be summarized as follows:

1. In the carrier (hemoglobin), a high affinity for oxygen in the presence of a moderate supply, and a lowered affinity in an oxygen-poor environment.
2. In the repository (myoglobin), a high affinity for oxygen under conditions where the carrier is giving up its supply.
3. In the carrier, the ability also to carry carbon dioxide away with it back to the lungs and to keep the pH of the tissues from falling.
4. In the carrier, a lessened affinity for oxygen at low pH, encouraging the transfer of oxygen at the tissues.

Most of these properties are illustrated in the saturation curves shown to the right below. The oxygen-binding curve for myoglobin is the hyperbola expected for simple one-to-one association of myoglobin heme and oxygen:

$$Mb + O_2 \rightleftharpoons MbO_2 \qquad K = \frac{[MbO_2]}{[Mb]\,[O_2]} = \begin{array}{l}\text{equilibrium}\\ \text{saturation}\\ \text{constant}\end{array}$$

If y is the fraction of myoglobin molecules saturated and if the oxygen concentration is expressed in terms of the partial pressure of oxygen, then

$$K = \frac{y}{(1-y)P_{O_2}} \qquad y = \frac{KP_{O_2}}{1+KP_{O_2}}$$

This is the equation of the hyperbola to the right labeled myoglobin.

Hemoglobin behaves differently. Its S-shaped, or sigmoid, curve requires an association-constant expression with a greater-than-first-power dependence upon the oxygen concentration:

$$K = \frac{[HbO_2]}{[Hb]\,[O_2]^n} \qquad y = \frac{KP_{O_2}{}^n}{1+KP_{O_2}{}^n}$$

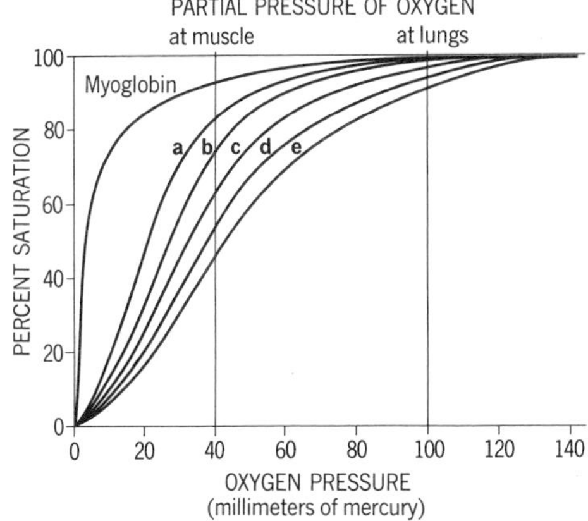

Oxygen binding curves for myoglobin and for hemoglobin at (a) *pH 7.6,* (b) *pH 7.4,* (c) *7.2* (d) *7.0, and* (e) *6.8. The tendency to release oxygen at low pH is known as the Bohr effect.*

The fact that n is observed to be nearer to 2.8 instead of 1.0 indicates that the binding of oxygen molecules to the four hemes is not independent, and that binding to any one heme is affected by the state of the other three. The first oxygen attaches itself very slowly to the heme. But the second and third bind more and more readily, and the fourth oxygen molecule binds several hundred times more rapidly than the first. Conversely, the sigmoid shape of the binding curve means that the molecule can release more oxygen in a low-oxygen environment than would be the case with a hyperbolic curve.

Oxyhemoglobin, with four bound oxygen molecules, behaves as a stronger acid than oxygen-free deoxyhemoglobin and loses a proton more readily. Hence, by Le Châtelier's principle, in an acid environment the oxygenation equilibrium should be shifted toward the deoxy-form and the removal of protons by binding. As the medium becomes more basic, the oxygenated form is favored and protons are liberated. This behavior is seen in the shift of the binding curve to the right with increasing acidity. The practical result of this behavior is that lactic and carbonic acids at the tissue reduce the pH and encourage the hemoglobin to drop its oxygen. The transportation of CO_2 back to the lungs is not a simple process, as can be seen in the diagram to the left.

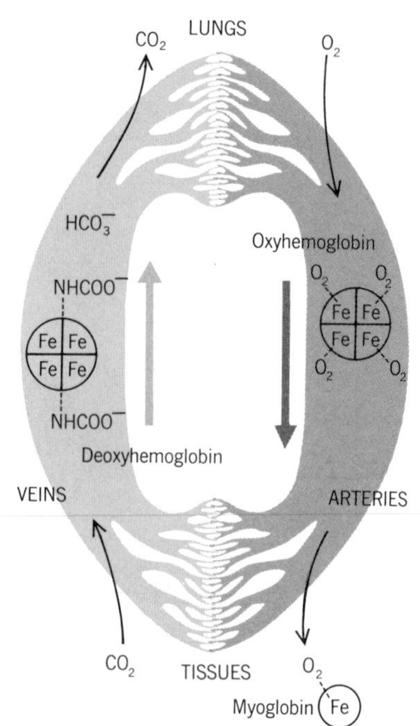

Oxygen is carried from the lungs to the tissues by hemoglobin, and carbon dioxide is carried back to the lungs. Part of the CO_2 transport occurs because the aliphatic amino groups can bind CO_2 directly to form carbamino compounds:

$$R\text{—}NH_2 + CO_2 = R\text{—}NHCOO^- + H^+$$

Another factor is the simple buffering ability of the deoxyhemoglobin molecule, which can combine with protons and carry CO_2 along in solution as bicarbonate ions.

In the fetus before birth, another transfer of oxygen has to take place—from the blood of the mother to that of the fetus. This means that the hemoglobin of the fetus should bind oxygen at concentrations where that of the mother is giving up oxygen, and must be more like myoglobin in its affinity for oxygen. Human fetal hemoglobin has the two α chains of the adult, but has its β chains replaced by two closely related γ chains. At pH 6.8, when the adult hemoglobin binding curve is e in the graph on page 45, that of the fetus would be close to curve b. The oxygen transfer is a three-step process: from the lungs to the hemoglobin of the mother, from there to that of the fetus, and then to the myoglobin of the fetus for storage until needed. Within 4–6 months after birth the last fetal hemoglobin disappears and is replaced by the adult form.

Myoglobin and hemoglobin were the first two globular proteins to have their detailed atomic structures worked out by x-ray diffraction, an achievement that earned the Nobel Prize in chemistry for M. F. Perutz and J. C. Kendrew in 1962 (4, 5). The high-resolution picture of myoglobin was completed in 1959 and has been refined and interpreted since then. The oxyhemoglobin structure was first deduced from a low-resolution analysis in 1960 by analogy with myoglobin and then confirmed at high resolution in 1967. The deoxyhemoglobin analysis, although only at low resolution so far, has already yielded dramatic results (page 56). And finally, the study of an insect hemoglobin with one chain has shown the surprising conservatism of the evolutionary process.

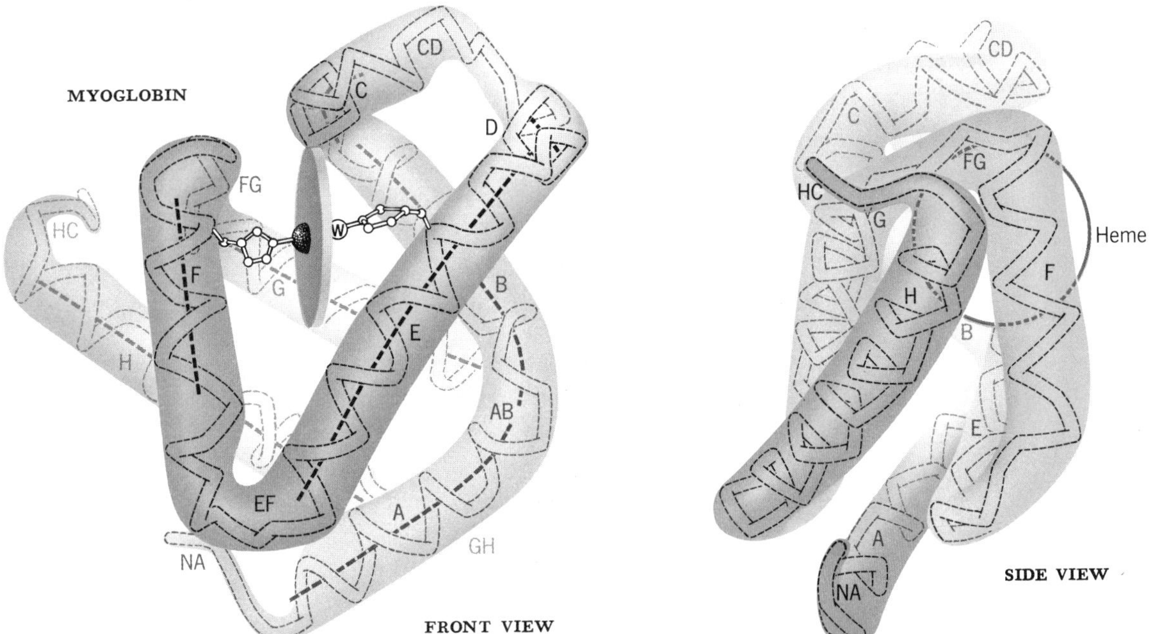

The myoglobin molecule is built up from eight stretches of α helix that form a box for the heme group. Histidines interact with the heme to the left and right, and the oxygen molecule sits at W. Helices E and F form the walls of a box for the heme; B, G, and H are the floor; and the C D corner closes the open end. The course of the main chain is outlined inside the color envelopes.

Helix	Length	Nonhelix	Length
		NA	2
A	16	AB	1
B	16		
C	7	CD	8
D	7		
E_1	10		
E_2	10	EF	8
F	10*	FG	4*
G	19	GH	5*
H	26*	HC	4*
Total length	121	....	32

** From their patterns of hydrogen bonding, FG1, GH6, and HC1 should be considered part of the helices to which they are connected.*

We have seen from this section that the list of chemical properties to be explained by the molecular structure is long. It would be pleasant to be able to provide complete explanations in the sections to follow, but this is not yet possible. We are only beginning to understand how the molecules might function, and much remains to be done. With these warnings, let us see what *is* known about the structures of these heme proteins.

3.2 MYOGLOBIN, THE CONTAINER*

The richest sources of myoglobin are the muscles of aquatic diving vertebrates such as seals, whale, and porpoises, and it was from sperm whale that Kendrew obtained myoglobin for the first protein x-ray structure analysis. The remarkable feature about the myoglobin molecule is that it turned out essentially to be a box for the heme group, built up from eight connected pieces of α helix. In the drawings above, the helices are lettered A through H starting from the amino terminal end. All the helices are right-handed, and they range in length from 7 residues in helices C and D to 26 in helix H. In all, 121 of the 153 residues are in helical regions. The molecule is an oblate spheroid, approximately 44 by 44 by 25 Å in size. The side view above shows how shallow the "box" really is.

* More on the myoglobin structure analysis can be found in references (6–10).

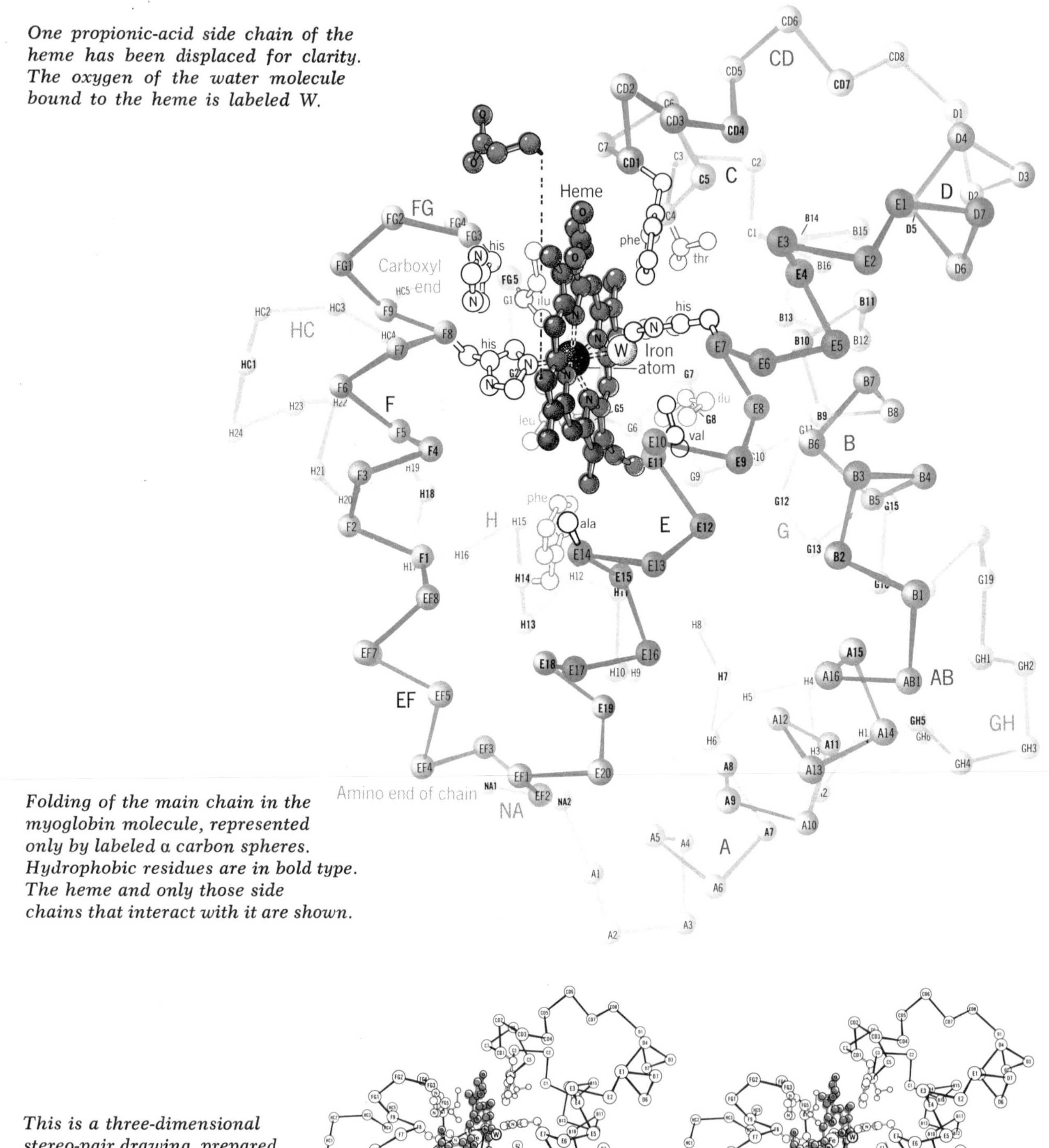

One propionic-acid side chain of the heme has been displaced for clarity. The oxygen of the water molecule bound to the heme is labeled W.

Folding of the main chain in the myoglobin molecule, represented only by labeled α carbon spheres. Hydrophobic residues are in bold type. The heme and only those side chains that interact with it are shown.

This is a three-dimensional stereo-pair drawing, prepared from atomic coordinates by courtesy of Dr. J. C. Kendrew, Cambridge, England. For information on viewing glasses, see the back cover.

It is impossible to build up a globular molecule from α helices unless they are folded back on top of one another, and between these eight helices the polypeptide chain is forced to bend. In some cases the transition from one helix to the next is abrupt, with no nonhelical residues, as from B to C or from D to E. There are gradual bends between F and G and between G and H. There are also two unhelical regions that seem to be important in their own right and not just as hinges between helices: chain CD and the corner of the all-important heme pocket, EF. There are short nonhelical "tails" at both ends of the chain, NA and HC. The presence of a Pro residue appears to be a sufficient but not a necessary condition for producing a bend. The sequence of myoglobin on page 52 shows a Pro at the beginning of helices C, F, and G, and another in the GH bend. Anticipating the fact that the hemoglobin chains will each turn out to be folded like myoglobin, it is seen that in the sequences of horse and human hemoglobin chains on page 52, Pro turns up in one chain or another at the beginning of helices A, C, D, E, F, G, and H, and in the interhelix regions, CD, EF, and GH, but never in the middle or near the end of a helix. Pro can be tolerated at the beginning of an α helix (page 51) because the first three N-terminal residues of such a helix need not use their amide nitrogens to form hydrogen bonds. The fact that this amide nitrogen is occupied with the side chain in Pro is hence no handicap.

The *raison d'etre* of the myoglobin molecule is the heme group, to whose iron atom the oxygen molecule binds. The purpose of the heme and of the polypeptide chain around it is to keep the ferrous iron from being oxidized (metmyoglobin, with a ferric iron, does not bind oxygen), and to give the iron its special oxygen-storage properties. The eight helices of the molecule are folded back upon themselves to form a pocket for the heme, as the diagram across the page and the stereo drawing below it show. The heme is completely buried except for one edge—that which contains the two hydrophilic propionic acid groups. The hydrophilic or charged polar groups are spread uniformly over the outside surface of the molecule. The strongly hydrophobic side chains (Val, Leu, Ilu, Met, and Phe) appear on the "inside" of the molecule in two ways. Some of them line the inner surfaces of the α helices, where the helices pack against one another. Others form a hydrophobic lining of the pocket for the heme. If several successive hydrophobic groups are to be on the same side of an α helix, then they must be spaced three or four residues apart along the polypeptide chain. A good example of this is the H helix (right). A hydrophobic group appears every three or four residues, placed just right to face the neighboring A helix and to nest against the EF bend.

The "V" produced by helices E and F provides the principal walls of the heme box (page 47). Helices B, G, and H form the floor, and the open end is closed off by C and CD. The nonhelical

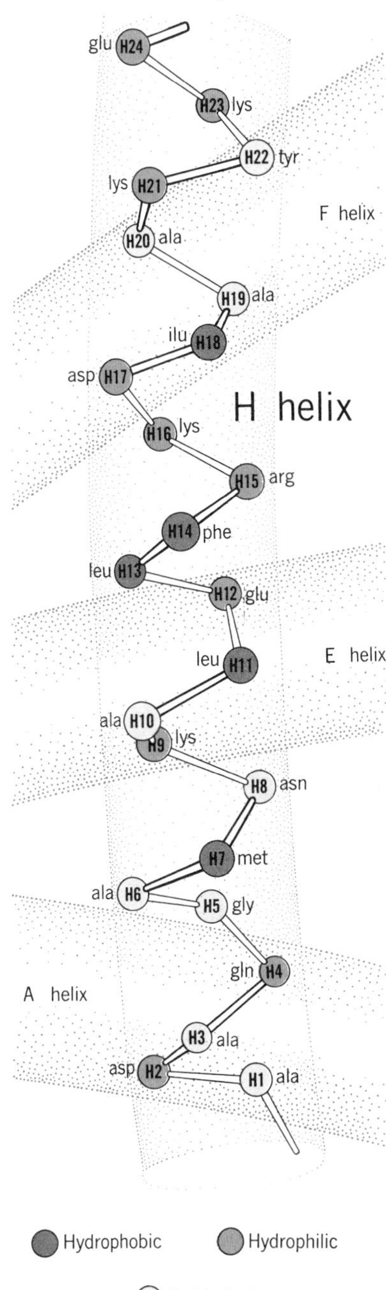

The H helix in myoglobin, showing the location of hydrophobic, ambivalent, and hydrophilic side chains. Note how the nearer surface of the H helix, which makes contact with the A, E, and F helices, is largely hydrophobic.

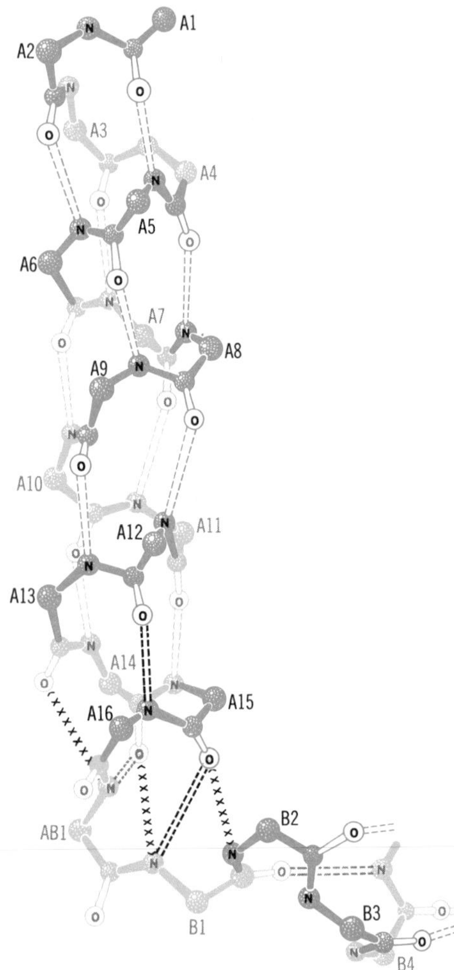

BENDING A CORNER BETWEEN THE A AND B HELICES

The last proper α helix hydrogen bond in the A helix is O_{12} . . . N_{16}. Thereafter, O_{13} does not form a hydrogen bond. O_{14} bonds to N_{AB1} and O_{15} bonds to N_{B1} as in a 3_{10} helix. O_{16} and O_{AB1} are unbonded, and O_{B1} points at right angles to the A helix axis, in the direction of the B helix axis.

======= *Hydrogen bonds*

xxxxxx *Formal hydrogen bonding pattern if the α helix continued beyond A_{16}*

CD region is itself important. It brings hydrophobic groups back again and again to line the inside wall of the heme box, by means of the sequence, starting from residue C3, of

-x-x-Leu-x-x-Phe-x-x-Phe-x-x-Leu-x-

The iron atom is octahedrally coordinated, as it usually is in inorganic complexes. Its six nearest-neighbor atoms, at the corner of the octahedron, each donate one electron pair to the orbitals around the iron. Four of these atoms are nitrogens in the porphyrin ring, as the sketch on page 44 shows. A fifth is one of the ring nitrogens of His F8, as shown on the right-hand page. This important histidine side chain is locked into position by a hydrogen bond from its other ring nitrogen to the carbonyl oxygen of Leu F4. All this scaffolding is designed to provide the possibility for the binding of the sixth octahedral ligand, which is the oxygen molecule itself. In oxymyoglobin it sits on the far side of the heme from His F8. In metmyoglobin, for which the most detailed structure analysis has been carried out, the oxygen site is occupied by a molecule of water, and it is the oxygen of this molecule that is shown as W in the drawing to the right. In deoxymyoglobin, x-ray analysis has shown the surprising result that the oxygen-binding site, rather than being filled with a water molecule, is simply empty.

On the far side of the oxygen molecule lies the side group of another histidine, His E7. It is too far from the iron to complex with it directly, but it does so indirectly through the oxygen or water molecule. Other small molecules or ions can also bind to myoglobin as the sixth iron ligand, including CO and NO for the ferro form and NO, OH^-, F^-, N_3^+, and H_2S for the ferri form. Carbon monoxide, to our detriment, forms a particularly strong bond. At equal concentrations of CO and O_2, there is 30–50 times as much binding of CO as O_2 in various myoglobins, and 120–550 times for different hemoglobins. In the presence of CO, O_2 is displaced and the molecule is finished as an oxygen carrier. The victim smothers.

The main building blocks of the molecule are the very stable α helices. But these helices are subject to deformations in many places, especially at their carboxyl-terminal ends, where the chain needs to be sent off in another direction. It is worth noting that, if polypeptide chains in proteins are synthesized from their N-terminal ends and if they then fold spontaneously, helix-*forming* side chains will be necessary at the N-terminal end of a nascent helix to initiate the folding, while helix-breaking groups will be needed at the *C-terminal* end. The A, C, E, and G helices all tighten up during their final turn of helix to form the hydrogen-bonding pattern of the 3_{10} helix (left), whereas the F and H helices loosen by one NH position to form something like a part of a turn of π helix. In neither case are these new helices stable enough to continue more than one turn, and the helicity of the polypeptide chain is brought to an end.

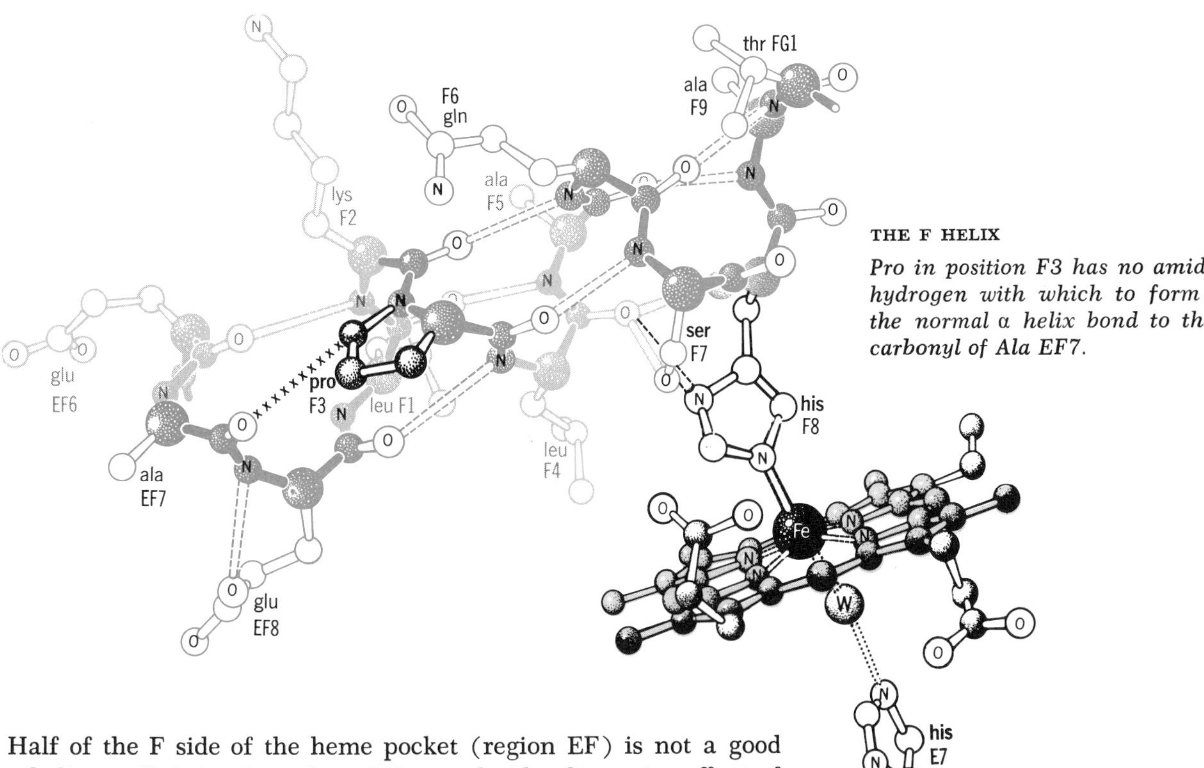

THE F HELIX

Pro in position F3 has no amide hydrogen with which to form the normal α helix bond to the carbonyl of Ala EF7.

THE HEME ENVIRONMENT

With a water or oxygen molecule present at W, the iron atom is octahedrally coordinated. The iron is slightly out of the heme plane in the direction of His F8. There are hydrogen bonds from the carbonyl of Leu F4 to the hydroxyl of Ser F7 and to the other ring nitrogen of His F8, which help to hold this His ring in the proper position for heme coordination.

Half of the F side of the heme pocket (region EF) is not a good α helix at all, being kept from being so by the disruptive effect of Pro F3 (above). The other side, the E helix, is regular but has a bend of 7° in the middle that helps it to bend around the heme and enclose the heme pocket.

The heme group is not covalently bound to the globin other than via this coordination complex. It can be removed gently from the globin in both myoglobin and hemoglobin, and can be put back to form a functioning molecule again. It is true that there are hydrogen bonds between the propionic acid side chains of the heme and side chains of the globin. The two acid groups extend out of the heme pocket into the surroundings and bend back to the main chain. The one that bends toward the F helix is hydrogen-bonded to a ring nitrogen of His FG3, which reaches out across the heme toward it. The other propionic acid chain is bonded to Arg CD3, which itself is also bonded to Asp E3. But the main stabilizing forces for the heme appear to be van der Waals or hydrophobic interactions (page 18). There are many hydrophobic groups packed around the heme: Ala E14, Val E11, Leu G5, Ilu G8, Phe H14, and CD1. The benzene ring of Phe CD1 is particularly close to the heme and is nearly parallel with it, suggesting an overlap of the π-electron clouds of the two resonating ring systems and possible electronic interaction. Other nonpolar groups in the CD bend and elsewhere cooperate to ensure that the entire heme pocket is strongly hydrophobic.

	NA 1	2	3	A 1	2	3	4	5	6	7	8	9	10	11	12	13	14	15	A 16	AB 1	B 1	2	3	4	5	6
MYOGLOBIN	val	.....	leu	ser	glu	gly	glu	trp	gln	leu	val	leu	his	val	trp	ala	lys	val	glu	ala	asp	val	ala	gly	his	gly
HEMOGLOBIN Horse α	val	.....	leu	ser	ala	ala	asp	lys	thr	asn	val	lys	ala	ala	trp	ser	lys	val	gly	gly	his	ala	gly	glu	tyr	gly
β	val	gln	leu	ser	gly	glu	glu	lys	ala	ala	val	leu	ala	leu	trp	asp	lys	val	asn	.....	.....	glu	glu	glu	val	gly
Human α	val	.....	leu	ser	pro	ala	asp	lys	thr	asn	val	lys	ala	ala	trp	gly	lys	val	gly	ala	his	ala	gly	glu	tyr	gly
β	val	his	leu	thr	pro	glu	glu	lys	ser	ala	val	thr	ala	leu	trp	gly	lys	val	asn	.....	.....	val	asp	glu	val	gly
γ	gly	his	phe	thr	glu	glu	asp	lys	ala	thr	ilu	thr	ser	leu	trp	gly	lys	val	asn	.....	.....	val	glu	asp	ala	gly
δ	val	his	leu	thr	pro	glu	glu	lys	thr	ala	val	asn	ala	leu	trp	gly	lys	val	asn	.....	.....	val	asp	ala	val	gly

	7	8	9	10	11	12	13	14	15	16	C 1	2	3	4	5	6	7	CD 1	2	3	4	5	6	7	8	D 1
MYOGLOBIN	gln	asp	ilu	leu	ilu	arg	leu	phe	lys	ser	his	pro	glu	thr	leu	glu	lys	phe	asp	arg	phe	lys	his	leu	lys	thr
HEMOGLOBIN Horse α	ala	glu	ala	leu	glu	arg	met	phe	leu	gly	phe	pro	thr	thr	lys	thr	tyr	phe	pro	his	phe	.....	asp	leu	ser	his
β	gly	glu	ala	leu	gly	arg	leu	leu	val	val	tyr	pro	trp	thr	gln	arg	phe	phe	asp	ser	phe	gly	asp	leu	ser	gly
Human α	ala	glu	ala	leu	glu	arg	met	phe	leu	ser	phe	pro	thr	thr	lys	thr	tyr	phe	pro	his	phe	.....	asp	leu	ser	his
β	gly	glu	ala	leu	gly	arg	leu	leu	val	val	tyr	pro	trp	thr	gln	arg	phe	phe	glu	ser	phe	gly	asp	leu	ser	thr
γ	gly	glu	thr	leu	gly	arg	leu	leu	val	val	tyr	pro	trp	thr	gln	arg	phe	phe	asp	ser	phe	gly	asn	leu	ser	ser
δ	gly	glu	ala	leu	gly	arg	leu	leu	val	val	tyr	pro	trp	thr	gln	arg	phe	phe	glu	ser	phe	gly	asp	leu	ser	ser

	2	3	4	5	6	7	E 1	2	3	4	5	6	7	8	9	10	11	12	13	14	E 15	16	17	18	19	20
MYOGLOBIN	glu	ala	glu	met	lys	ala	ser	glu	asp	leu	lys	lys	his	gly	val	thr	val	leu	thr	ala	leu	gly	ala	ilu	leu	lys
HEMOGLOBIN Horse α	.....	.....	.....	.....	.....	gly	ser	ala	gln	val	lys	ala	his	gly	lys	lys	val	ala	asp	gly	leu	thr	leu	ala	val	gly
β	pro	asp	ala	val	met	gly	asn	pro	lys	val	lys	ala	his	gly	lys	lys	val	leu	his	ser	phe	gly	glu	gly	val	his
Human α	.....	.....	.....	.....	.....	gly	ser	ala	gln	val	lys	gly	his	gly	lys	lys	val	ala	asp	ala	leu	thr	asn	ala	val	ala
β	pro	asp	ala	val	met	gly	asn	pro	lys	val	lys	ala	his	gly	lys	lys	val	leu	gly	ala	phe	ser	asp	gly	leu	ala
γ	ala	ser	ala	ilu	met	gly	asn	pro	lys	val	lys	ala	his	gly	lys	lys	val	leu	thr	ser	leu	gly	asp	ala	ilu	lys
δ	pro	asp	ala	val	met	gly	asn	pro	lys	val	lys	ala	his	gly	lys	lys	val	leu	gly	ala	phe	ser	asp	gly	leu	ala

	EF 1	2	3	4	5	6	7	8	F 1	2	3	4	F 5	6	7	8	9	FG 1	2	3	4	5	G 1	2	3	4
MYOGLOBIN	lys	lys	gly	his	his	glu	ala	glu	leu	lys	pro	leu	ala	gln	ser	his	ala	thr	lys	his	lys	ilu	pro	ilu	lys	tyr
HEMOGLOBIN Horse α	his	leu	asp	asp	leu	pro	gly	ala	leu	ser	asp	leu	ser	asn	leu	his	ala	his	lys	leu	arg	val	asp	pro	val	asn
β	his	leu	asp	asn	leu	lys	gly	thr	phe	ala	ala	leu	ser	glu	leu	his	cys	asp	lys	leu	his	val	asp	pro	glu	asn
Human α	his	val	asp	asp	met	pro	asn	ala	leu	ser	ala	leu	ser	asp	leu	his	ala	his	lys	leu	arg	val	asp	pro	val	asn
β	his	leu	asp	asn	leu	lys	gly	thr	phe	ala	thr	leu	ser	glu	leu	his	cys	asp	lys	leu	his	val	asp	pro	glu	asn
γ	his	leu	asp	asp	leu	lys	gly	thr	phe	ala	gln	leu	ser	glu	leu	his	cys	asp	lys	leu	his	val	asp	pro	glu	asn
δ	his	leu	asp	asn	leu	lys	gly	thr	phe	ser	gln	leu	ser	glu	leu	his	cys	asp	lys	leu	his	val	asp	pro	glu	asn

	5	6	7	8	G 9	10	11	12	13	14	15	16	17	18	19	GH 1	2	3	4	5	6	H 1	2	H 3	4	5
MYOGLOBIN	leu	glu	phe	ilu	ser	glu	ala	ilu	ilu	his	val	leu	his	ser	arg	his	pro	gly	asn	phe	gly	ala	asp	ala	gln	gly
HEMOGLOBIN Horse α	phe	lys	leu	leu	ser	his	cys	leu	leu	ser	thr	leu	ala	val	his	leu	pro	asn	asp	phe	thr	pro	ala	val	his	ala
β	phe	arg	leu	leu	gly	asn	val	leu	ala	leu	val	val	ala	arg	his	phe	gly	lys	asp	phe	thr	pro	glu	leu	gln	ala
Human α	phe	lys	leu	leu	ser	his	cys	leu	leu	val	thr	leu	ala	ala	his	leu	pro	ala	glu	phe	thr	pro	ala	val	his	ala
β	phe	arg	leu	leu	gly	asn	val	leu	val	cys	val	leu	ala	his	his	phe	gly	lys	glu	phe	thr	pro	pro	val	gln	ala
γ	phe	lys	leu	leu	gly	asn	val	leu	val	thr	val	leu	ala	ilu	his	phe	gly	lys	glu	phe	thr	pro	glu	val	gln	ala
δ	phe	arg	leu	leu	gly	asn	val	leu	val	cys	val	leu	ala	arg	asn	phe	gly	lys	glu	phe	thr	pro	gln	met	gln	ala

	6	7	8	9	10	11	12	13	14	15	16	17	18	19	20	H 21	22	23	24	HC 1	2	3	4	5
MYOGLOBIN	ala	met	asn	lys	ala	leu	glu	leu	phe	arg	lys	asp	ilu	ala	ala	lys	tyr	lys	glu	leu	gly	tyr	gln	gly
HEMOGLOBIN Horse α	ser	leu	asp	lys	phe	leu	ser	ser	val	ser	thr	val	leu	thr	ser	lys	tyr	arg						
β	ser	tyr	gln	lys	val	val	ala	gly	val	ala	asn	ala	leu	ala	his	lys	tyr	his						
Human α	ser	leu	asp	lys	phe	leu	ala	ser	val	ser	thr	val	leu	thr	ser	lys	tyr	arg						
β	ala	tyr	gln	lys	val	val	ala	gly	val	ala	asn	ala	leu	ala	his	lys	tyr	his						
γ	ser	trp	gln	lys	met	val	thr	gly	val	ala	ser	ala	leu	ser	ser	arg	tyr	his						
δ	ala	tyr	gln	lys	val	val	ala	gly	val	ala	asn	ala	leu	ala	his	lys	tyr	his						

SEQUENCES OF MYOGLOBIN AND HEMOGLOBIN CHAINS

The amino-acid sequences of sperm-whale myoglobin, horse hemoglobin α and β, and human α, β, γ, and δ. Residues that are identical in all the hemoglobin chains or in all seven chains are shown with a dark color tint, and those of the conservative substitution of one amino acid by a closely related one have a light color tint. Deletions in a chain are shown by dots. Sequences are listed, as always, from N-terminal to C-terminal end. The letter/number notation is that for myoglobin (10). The hemoglobin workers begin the H helix one residue earlier (17, 21), so that the invariant Tyr is H22 in myoglobin and H23 in hemoglobin.

There are relatively few places where one has the impression that *specific* side-chain interactions are designed to make the helices fold together in a particular way. The 11 Ser and Thr all form hydrogen bonds, but to main-chain carbonyl oxygens in the same α helix rather than from one helix to another. Many of the hydrogen bonds which have been found seem to be adventitious rather than vital. One possible exception is the evolutionarily conserved Tyr H22, which is hydrogen-bonded to the carbonyl group of Ilu FG5. Another is Thr C4, whose side-chain hydroxyl is bonded to the main-chain carbonyl of His C1. This cross-bridging, which helps to form the CD corner, will be frequently encountered in Thr and Ser in less α-helical proteins. But as with the heme-globin interaction, the main stabilizing force for the folding of the myoglobin molecule seems to be van der Waals forces or hydrophobic interactions. The helices fold against one another in the proper way because they prefer to match "oily" sides and to remove their nonpolar side chains from the aqueous environment. The proper folding is aided by cues such as the rare side-chain interactions of Tyr H22, Thr C4, Arg CD3/Asp E3, and a few others.

The table on the opposite page gives the amino acid sequences of sperm whale myoglobin; horse hemoglobin α and β chains; and human hemoglobin α, β, γ, and δ. Anticipating the observation that myoglobin and all the hemoglobin chains will be found to have the same folding, we can use these comparisons between chain types and species to learn which residues are absolutely essential for the folding and operation of a myoglobin/hemoglobin type of molecule. The myoglobin sequence has been solved completely for sperm whale (11), porpoise and seal (12), and partially for humpback whale, dolphin, and horse (11). The hemoglobin α sequence is known for man, gorilla, horse, donkey, cow, sheep, llama, pig, rabbit, mouse, and carp. The β sequence is known for all the above plus camel, goat, and lemur, and the γ and δ sequences for man. The sequence of the one-chain molecule of lamprey globin is also known. The absolutely constant residues in all the above chains are only seven in number: Gly B6, Phe CD1, His E7, Leu F4, His F8, Lys H9, and Tyr H22. In addition, Pro C2 is Thr only in pig α chain, Gly E6 is Ala only in lamprey globin, and Thr C4 is Gly in dolphin myoglobin and Ala in lamprey globin. The reasons for most of these are apparent: Phe CD1, His E7, Leu F4, and His F8 all interact with the heme group. Gly B6 and E6 occur at the close contact of the B and E helices, where there is no room for a side group. Thr C4 and Tyr H22 help to pull the molecule into shape, and Pro C2 helps to form the BC corner. But the purpose of Lys H9, which appears to point out away from the molecule into its surroundings, is a mystery. What importance does it have that it should be preserved so carefully through 500 million years of evolution?

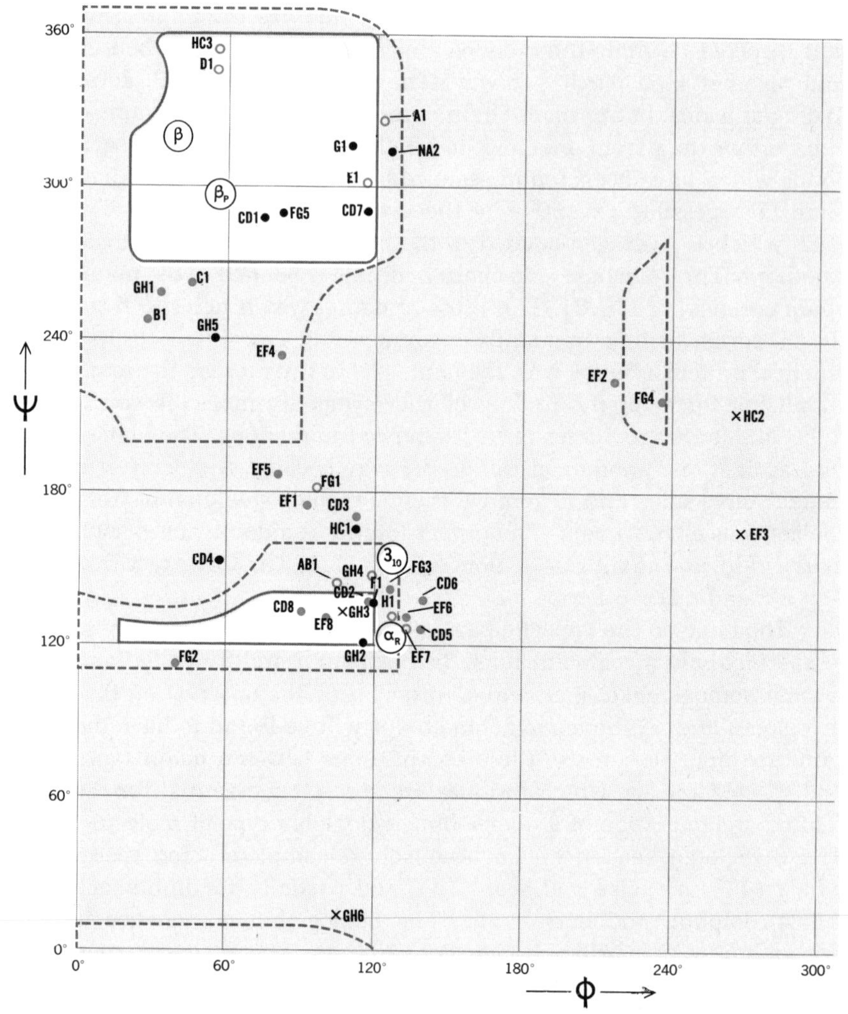

Ramachandran plot of the non-α-helical regions in myoglobin. Black dots denote hydrophobic residues; solid color dots, hydrophilic; open color dots, ambivalent. Crosses denote gly. The first α carbon in a helix will not necessarily have the (φ, ψ) value of an ideal helix.

The folding of the myoglobin chain can be summarized by the Ramachandran plot above, which has the idealized α and 3_{10} helices marked, and all the nonhelical residues of myoglobin. Note the tendency even of nonhelical α-carbons to adopt configurations close to that of one of these helices, the secondary predominance of β-like configurations, and the apparently only weakly forbidden region between them across the $\psi = 180°$ line.

The suggestion has been made several times that myoglobin and hemoglobin were related in structure. To what extent is this true, and how is hemoglobin folded?

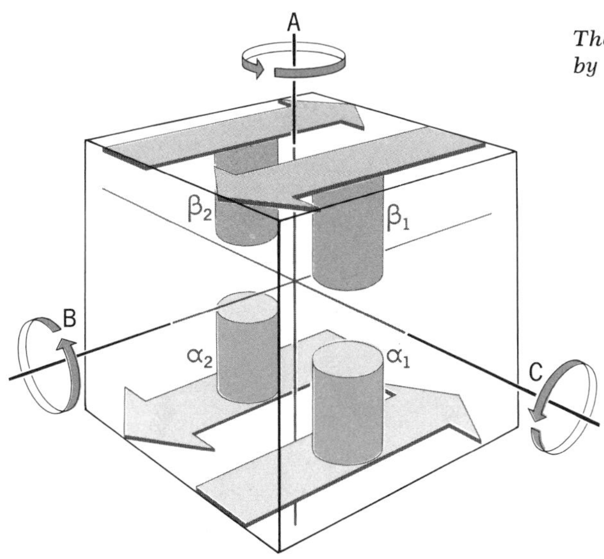

The symmetry of the hemoglobin molecule is shown by these four arbitrary asymmetric objects.

Result of 180° rotation about:

A AXIS

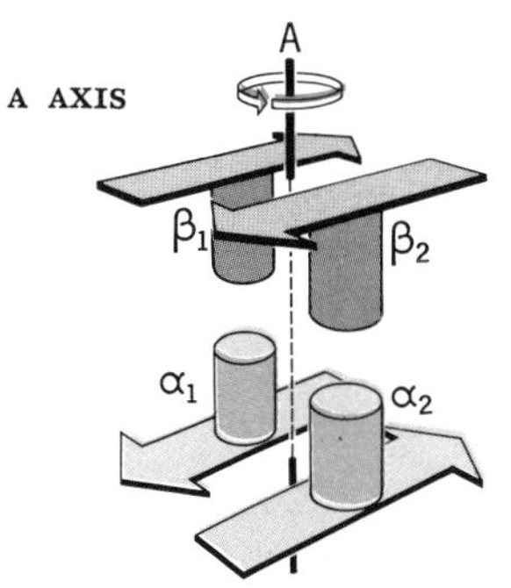

B AXIS

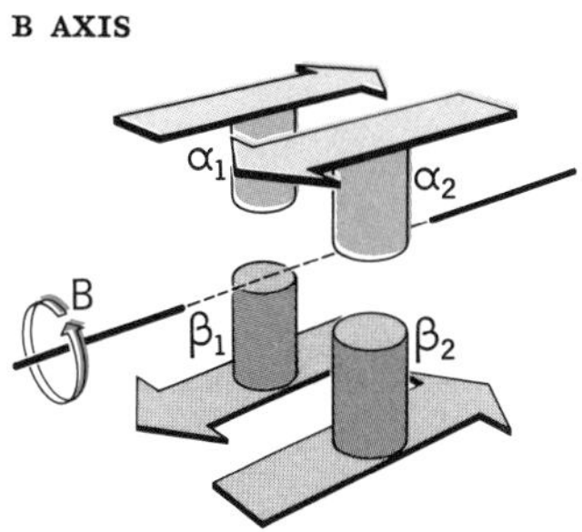

C AXIS

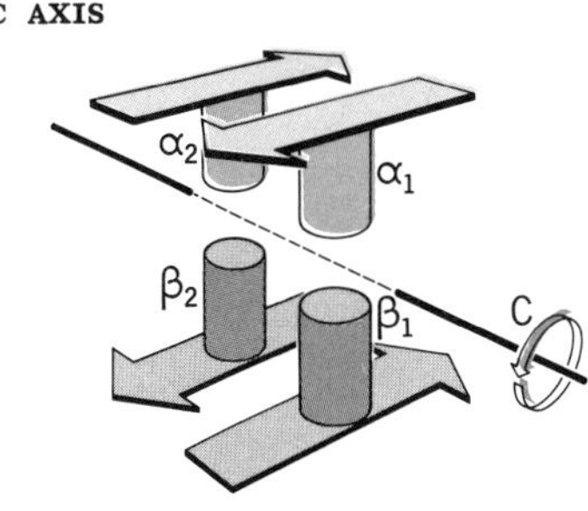

With two of type α and two of β, symmetry axis A is a true twofold axis, whereas B and C are both pseudo-axes of symmetry.

3·3 HEMOGLOBIN, THE CARRIER*

Hemoglobin had been known for a long time on chemical grounds to be made up of two α and two β chains. From the symmetry of the crystals, Perutz deduced as early as 1938 that one α and β pair was related to the other by a twofold axis (24)—that is, if one pair were rotated by 180° about a particular line, then they would superimpose on the other. But it was not until 1960, when the first low-resolution x-ray structure analysis was completed, that it was realized that the α and β chains themselves were folded in a similar way, and that both were like myoglobin. The complete hemoglobin molecule nearly has what is called 222 symmetry, with three mutually perpendicular twofold axes intersecting at the center of the molecule.

The meaning of 222 symmetry is shown at the top of the page. The subunits, arrows with cylinders attached, bear no resemblance to chains of hemoglobin, but the symmetry relationships are the same. If the entire assembly is given a 180° rotation about the vertical twofold axis, marked A, then subunit α_1 superimposes on α_2, and β_1 on β_2. Axis B would map subunit α_1 onto β_2, and α_2 onto β_1. But if the α and β subunits are different in detail, as they are in hemoglobin, then B is only a pseudo-symmetry axis. Similarly, axis C is also a pseudo axis of symmetry, for α_1 and β_1 are not identical in hemoglobin, nor are α_2 and β_2.

* The full hemoglobin story is to be found in references (13–23). Extensive sequence tabulations, for the globins and many other proteins, are available in Dayhoff and Eck (11).

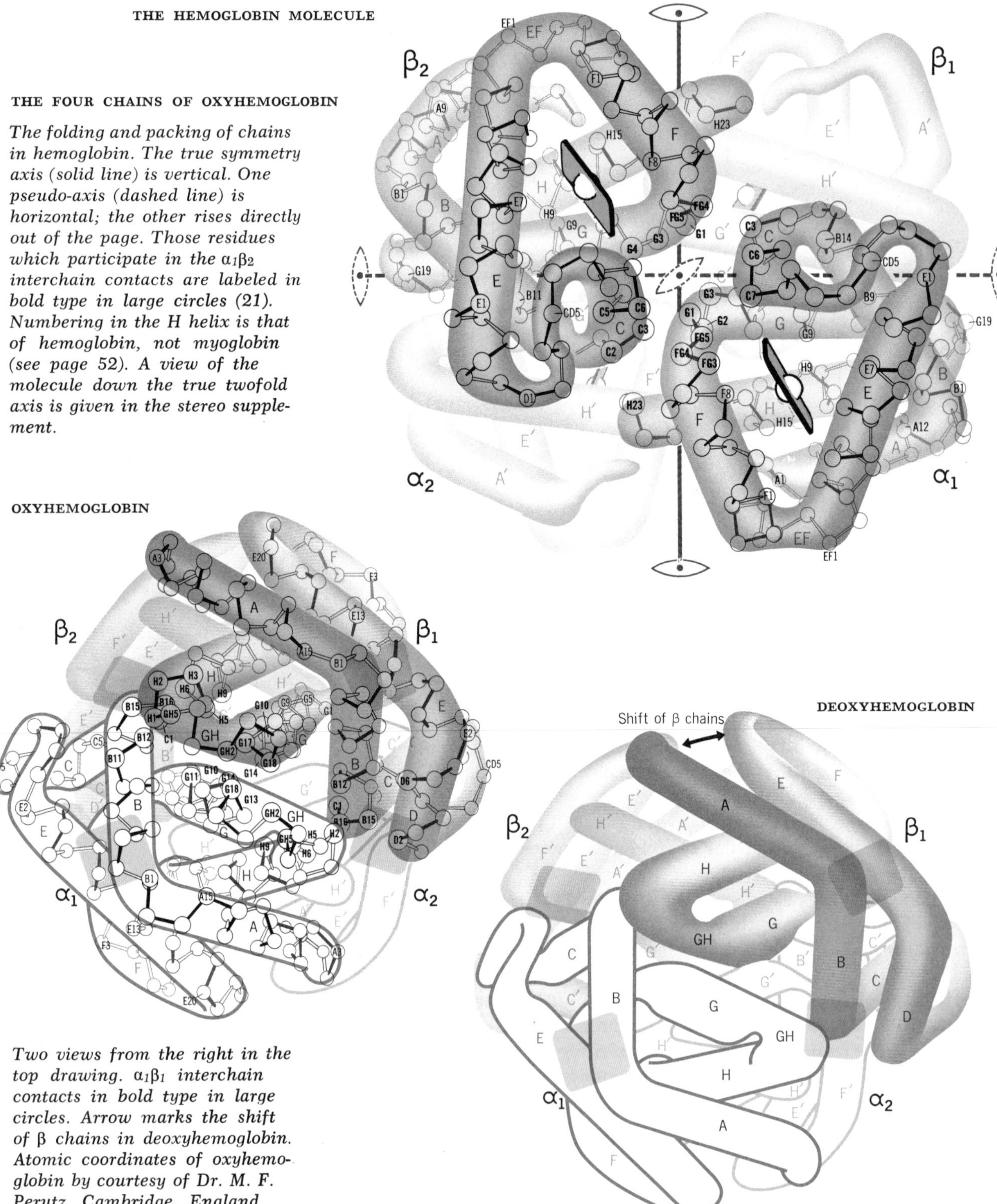

THE FOUR CHAINS OF OXYHEMOGLOBIN

The folding and packing of chains in hemoglobin. The true symmetry axis (solid line) is vertical. One pseudo-axis (dashed line) is horizontal; the other rises directly out of the page. Those residues which participate in the $\alpha_1\beta_2$ interchain contacts are labeled in bold type in large circles (21). Numbering in the H helix is that of hemoglobin, not myoglobin (see page 52). A view of the molecule down the true twofold axis is given in the stereo supplement.

OXYHEMOGLOBIN

DEOXYHEMOGLOBIN

Two views from the right in the top drawing. $\alpha_1\beta_1$ interchain contacts in bold type in large circles. Arrow marks the shift of β chains in deoxyhemoglobin. Atomic coordinates of oxyhemoglobin by courtesy of Dr. M. F. Perutz, Cambridge, England.

The actual arrangement of the four chains of horse oxyhemoglobin is shown at the top and middle of the opposite page. The molecule is roughly spherical, 64 by 55 by 50 Å. The four heme pockets, easily identifiable by their V-shaped E/F sides, are all exposed at the surface of the molecule. The heme groups of chains α_2 and β_1 are particularly close, as are those of α_1 and β_2. (The subscripts to α and β merely identify relative positions in the molecule and have no chemical significance.)

Each of the chains is folded in much the same way as the myoglobin molecule, with only minor differences. They both have the essential His residues at E7 and F8 and the other constant residues mentioned in connection with myoglobin. The chief difference is the lack of a D helix in the α chain, as shown to the right. Myoglobin has 153 residues; the α chain has 12 fewer. Six of these disappear with the missing D helix, and the other 6 are absent from the carboxyl end of the H helix. The β chain has 7 fewer residues than myoglobin: 6 because of a shorter H helix, as in the α chain, 1 extra at the N-terminal end of the chain, and 2 missing from the AB corner. There are several differences in details of helix structure. The final turn of the E helix is irregular in the β chain, and the entire E helix is irregular in the α chain, with abnormal hydrogen bonding. The short C helix, which was a somewhat distorted α helix in myoglobin, is a 3_{10} helix in both hemoglobin chains. The three chains are definitely variations on a theme, but the theme itself is still plainly visible.

The packing of chains into the hemoglobin molecule is such that there is close, interlocking contact of side groups between unlike chains but virtually no contact between α and α, or β and β. There are two kinds of unlike-chain contacts: between chains with neighboring hemes ($\alpha_1\beta_2$ or $\alpha_2\beta_1$), and between chains with widely separated hemes ($\alpha_1\beta_1$ or $\alpha_2\beta_2$). There are a few hydrogen bonds and charged-group interactions, but the great majority of contact interactions are hydrophobic. As with the heme and globin interactions and the packing of helices in myoglobin, the packing of chains in hemoglobin is largely a matter of entropy.

The $\alpha_1\beta_1$ contact is more extensive than the $\alpha_1\beta_2$ type, with about 34 side chains being involved as compared with 19. The region involved in $\alpha_1\beta_1$ contact is shown in the middle drawing on the left-hand page and involves the residues from G10 to H9, along with parts of the B and D helices of the β chain to the right and the B helix of the α chain on the left. The $\alpha_1\beta_2$ contact region is shown in the top drawing and consists mainly of the C helices and the FG bends.

It is interesting to compare the amino acid residues at these chain contact points, which are inside the molecule in hemoglobin but are on the outer surface of myoglobin. Pro D2 and Met D6 in the β chain, for example, are closely packed against hydrophobic

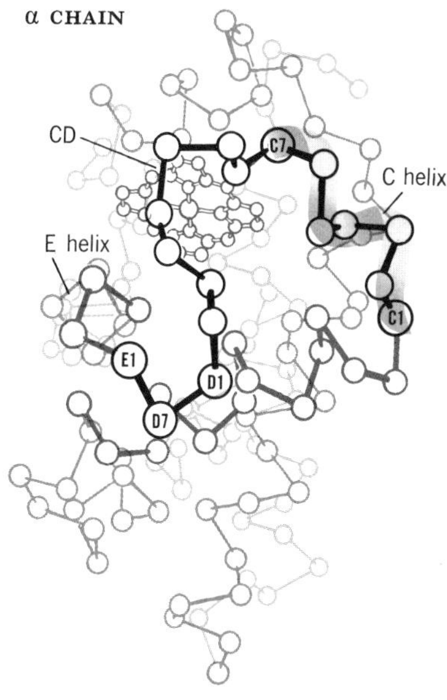

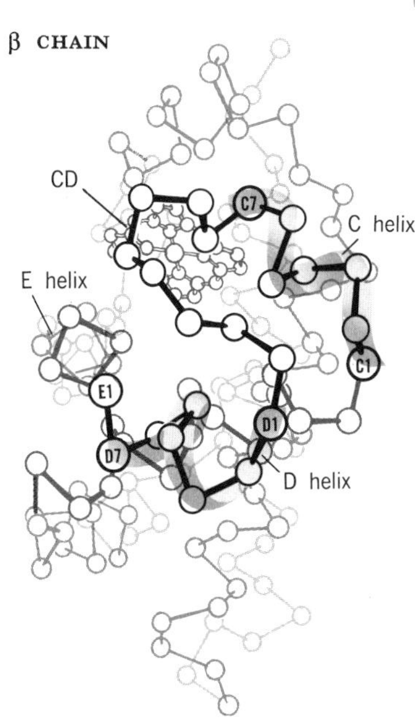

The α chain (top) lacks the six residues which in the β chain and in myoglobin form the D helix. The course of the polypeptide chains from C1 to E1 is shown in black. The C and D helices are overprinted in color.

groups in an α chain. When all the hemoglobin sequences mentioned earlier in connection with myoglobin are examined, it is found that site D2 is only found as Pro or Ala, and D6 is either Met or Leu, both cases of the conservative substitutions of one nonpolar side chain for another. But in the one-chain lamprey hemoglobin, sites D2 and D6 are Ala and Lys, and in myoglobin both sites are charged, Glu and Lys. The replacement of hydrophobic groups by charged hydrophilic ones makes these surfaces of the molecule prefer an aqueous environment and helps to prevent the formation of a dimer or tetramer. At several other contact points, nonpolar groups in hemoglobin are found to be replaced by Lys or other charged polar groups in myoglobin.

It is hardly to be expected that such strangely shaped objects as the hemoglobin chains would pack together perfectly, and they do not. Running down the true twofold axis for the entire 50-Å diameter of the molecule is a channel, 20 Å wide, which separates each pair of like chains by 10 Å at the top and bottom of the molecule. The channel is lined with polar groups, and for reasons which are about to be mentioned, may play some role in producing the pH dependence of oxygen binding, or the Bohr effect.

The molecule described so far is that of oxyhemoglobin, with four bound oxygen molecules. In deoxyhemoglobin the folding of the chains appears to be virtually the same, but their arrangement is different (19). An exact description of the relative motion of the four chains upon the addition or removal of oxygen would be difficult and hard to visualize. But it is approximately true that the $\alpha_1\beta_1$ and $\alpha_2\beta_2$ halves move as rigid units and that these two units slide past one another, as shown in the middle and bottom of page 56. The separation of hemes in unlike chains alters very little—the distances between iron atoms are shown to the left. But the two α hemes come about 1 Å closer together, and, most importantly, the two β hemes separate by 6.5 Å.

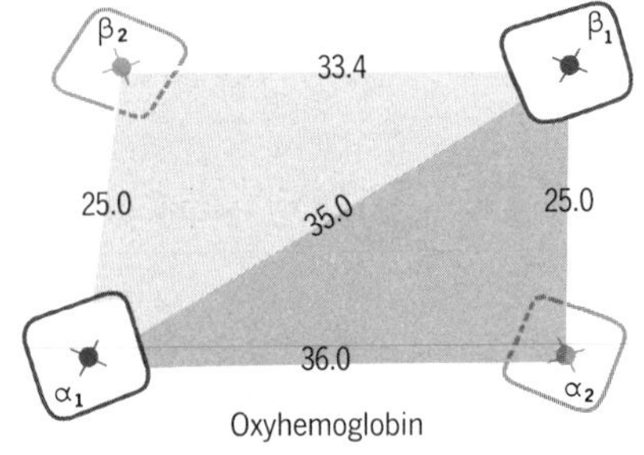

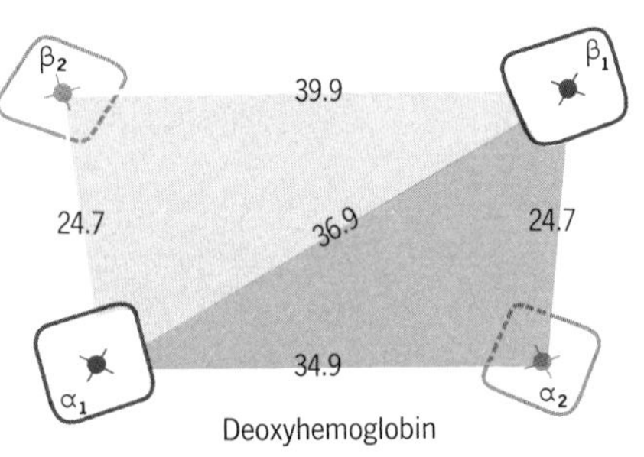

Relative motion of α and β chains in the two states of hemoglobin. Distances between iron atoms are in Ångstrom units. The hemes form a rough tetrahedron.

The hemoglobin molecule is seen to be, not an inert receptacle, but a functioning piece of machinery of molecular dimensions. But does this movement of the chains offer any explanation of the peculiar chemical properties of hemoglobin? It is known that if hemoglobin is broken down into dimers or monomers, both the sigmoid binding curve and the Bohr effect disappear. How do the four chains as a unit produce these phenomena?

The answers have been suggested, but not yet proved, by x-ray analysis. The heme/heme interaction that produces the all-or-nothing binding behavior almost surely occurs between the closely spaced hemes of the α_1 and β_2 chains and the α_2 and β_1 chains. A six-residue connection between these hemes may be involved:

Heme(α)· · ·Phe CD1(α)--Thr C6(α)--Tyr C7(α)· ·
· Arg C6(β)--Phe C7(β)--Phe CD1(β)· · ·Heme(β)

Not only the Heme···Phe CD1 interactions, but the Tyr···Arg interchain contact as well, seem to involve an overlap of electron clouds in resonating systems and π–π electronic interaction. The malfuctioning of the hemoglobin mutant with Ser in place of Phe CD1 (right) emphasizes the importance of this residue (25).

It is to be expected that a mechanism will be worked out to relate the oxygenation of the first heme, the conformational change, and the ease of oxygenation of the other hemes in the molecule. But what about the observation that the deoxygenated hemoglobin binds protons more readily than the oxygen-bearing form and hence is preferred in an acid environment? Again, the x-ray results are too new for definite conclusions to be drawn, but one possibility may be that the central cavity is involved and that the movement of the chains uncovers previously buried basic groups that can then bind protons.

We are now in the position of New Guinea natives who have been told in detail what an automobile does, have seen how the pistons of an internal combustion engine move up and down, but are not quite sure yet what the connection between these two sets of information may be. But now, for the first time, we know that it is possible to work out a molecular explanation for protein behavior, and that the possibilities are limited only by our own ingenuity.

3.4 THE HISTORY OF A PROTEIN: MOLECULAR EVOLUTION*

That myoglobin and the chains of hemoglobin are folded alike is no accident; they are cousins. One fact that has emerged in the last 15 years from the perfection of protein sequence and structure analysis methods is that every globular protein carries a record of its history in its structure. By studying the amino acid sequences of equivalent proteins from different species or of related proteins from the same species, we can pin down with great precision the relationships between the species and how the proteins evolved. What is more, the three-dimensional folding of two proteins may show an evolutionary connection that is no longer easy to recognize from the sequences alone.

Suppose that we were given only the comparative sequence information on page 52. What could we learn about the history of the globins? The first observations are that, considering only chain length, myoglobin appears to be in a class by itself, both human and horse α are alike, and all the other β, γ, and δ chains form a third class. Next, a detailed comparison of the differences in se-

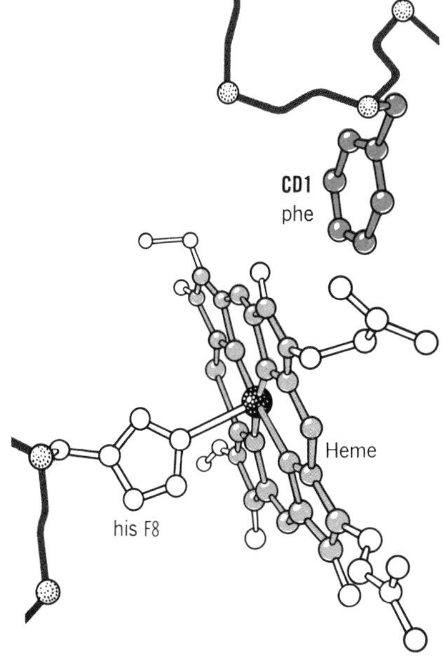

All normal hemoglobins and myoglobin have a Phe at CD1 which is close to and nearly parallel to the heme. There is probably π-orbital electronic interaction, possibly as part of the heme-heme interaction in hemoglobin. Hemoglobin Hammersmith, with Ser instead of Phe, is pathological (20).

* For more on protein evolution, see references (26–33).

DIFFERENCES IN AMINO ACID SEQUENCES

	Horse α	Human α	Horse β	Human β	Human δ	Human γ	Whale Mb
Total Residues:	141	141	146	146	146	146	153
Horse α	0	18	84	86	87	87	118
Human α	18	0	87	84	85	89	115
Horse β	84	87	0	25	26	39	119
Human β	86	84	25	0	10	39	117
Human δ	87	85	26	10	0	41	118
Human γ	87	89	39	39	41	0	121
Whale Mb	118	115	119	117	118	121	0

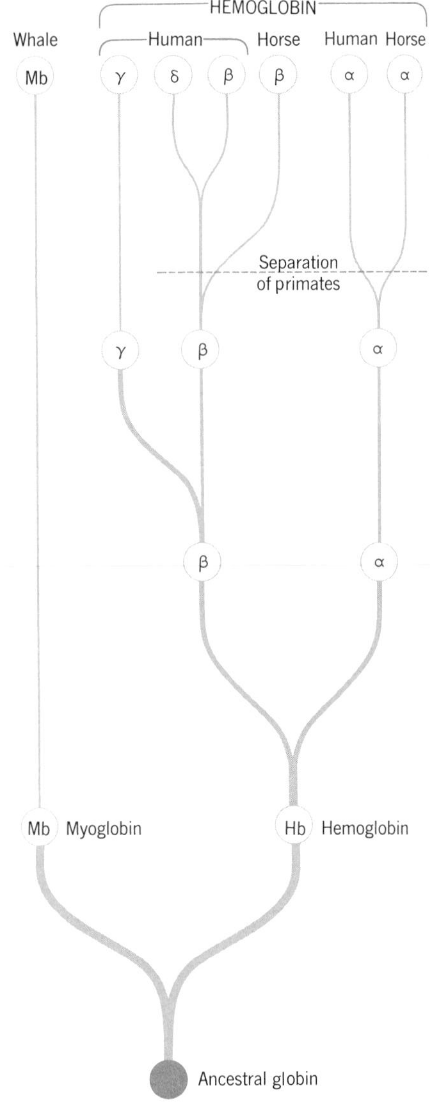

THE EVOLUTION OF THE GLOBINS

quences would produce the table of globin evolution shown above. (In obtaining these numbers, the sequences were aligned as on page 52, and gaps or absences were treated formally as a special twenty-first kind of residue.) It is immediately obvious that human β and δ chains are the most nearly alike and that horse and human α are the next most similar pair. Human β and δ are both equally different from horse β, and all the β and δ chains are about equally different from human γ. The **β**, γ, and δ chains as a class are equally far removed from the α, and all the hemoglobin chains, no matter of what type or which species, are equally far removed from whale myoglobin.

The only rational interpretation that can be placed on these data is shown in the tree to the left. The inferred sequence of events is the following. At some time in the far past, a primitive common ancestor of whales, horses, and men had only one oxygen-binding heme protein, which probably resembled myoglobin in general outline. This protein was coded for by one particular gene. The vast majority of random mutations in this gene, producing alterations in amino acid sequence, would be lethal, and, in general, only mutations that led to conservative substitutions of amino acids could be tolerated. Changes in sequence with time over the generations would be slow (33).

At some point in the history of the species, this particular gene doubled or duplicated (by ways that are familiar to the geneticist and are discussed thoroughly by Dixon [29]). With two genes possibly making the same protein, there were two improvements: A mutation in only one of these genes would now not be fatal, and the two genes could then evolve slowly along independent paths. One protein gradually became more suited to oxygen transport and the other to storage. These proteins could now be called "hemoglobin" and "myoglobin."

At a later time the hemoglobin gene doubled again and diverged with the production of α and β chains. Developing along with this divergence and making it advantageous to the organism was the buildup of four chains into a more efficient oxygen-carrying unit. This second doubling, incidentally, occurred some time after the lineage of the primitive lamprey diverged from ours, but before that of carp and the higher fish, for the lamprey has a one-chain hemoglobin, whereas carp hemoglobin is the familiar $\alpha_2\beta_2$ variety.

Later still, the β gene diverged again into β and γ, and, at a somewhat later date, the primates branched off from the rest of the mammals and separated the ultimate lineage of horse and man. A horse γ sequence is not available, so it might be possible to claim a reversal of the order of these last two branchings and to maintain that the β/γ split occurred in the course of primate evolution. However, the evidence from the table that horse and human β are closer than human β and γ makes this unlikely. Finally, quite late in the history of the primates, there was a last doubling of the β gene again and the appearance of the human δ.

The result of this process in man is the presence of four proteins, each tailored for the role it plays: a monomeric oxygen-storage protein, an adult oxygen-transport protein with the structure $\alpha_2\beta_2$, a special $\alpha_2\gamma_2$ fetal protein with greater oxygen affinity, and a new minor-component protein of unknown use, $\alpha_2\delta_2$. The suggestion that some of this minor protein's properties are better than those of the standard adult form raises the possibility that we may be observing the rise of the eventual successor to $\alpha_2\beta_2$ hemoglobin.

The globin record is complicated by the fact that one must work with divergences in species and in chain types simultaneously. For a really good map of the evolutionary history of living organisms, one should select a nearly universal protein but one that has not split up into subvariants. Such a molecule will be discussed in the next few pages. But dramatic confirmation of these ideas has recently come from x-ray analyses of the globins of the larva of a fly (23) and a marine annelid worm (34).

The larva of *Chironomus thummi* has a red, oxygen-binding heme protein about the size of myoglobin, which disappears in the adult. It was originally called an erythrocruorin and was classified with the large collection of various invertebrate oxygen-binding proteins. But the x-ray analysis (23) showed that in reality it was an insect hemoglobin (page 62, top). It, too, has a heme group in a V-shaped pocket formed by two of eight helices, and all eight are arranged in the same way as in myoglobin.

The same "myoglobin fold" has turned up again in the annelid worm *Glycera dibranchiata* (34), and raises a puzzling problem (page 62, bottom). Hemoglobins occur sporadically among the invertebrate phyla, in no obvious pattern. Does this distribution represent descent of these species from a common hemoglobin-

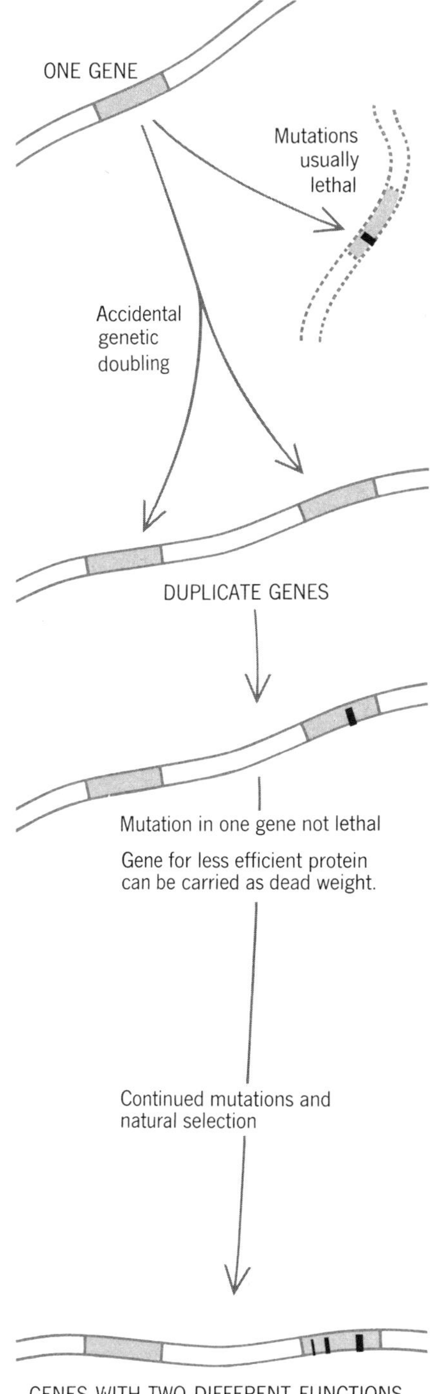

The process of gene doubling, mutation, and evolution.

bearing ancestor, or the repeated development of an oxygen-binding heme protein in the face of similar environmental problems and similar selection pressures? It is hard to see a common line of descent snaking in so unsystematic a way through so many different phyla, and the "repeated evolution" or "convergence on function" argument seemed plausible as long as the criteria for calling these molecules "hemoglobins" were only oxygen-binding properties and heme spectra. But now, is it possible to believe that the entire eight-helix folding pattern of the myoglobin molecule is so essential for its operation that the fold would re-evolve independently many times? It seems more probable that the myoglobin fold developed in oxygen-binding proteins *before* the divergence of annelids, insects, and what were to become chordates, and that those species which now lack hemoglobin have lost it rather than never having had it. Much detective work remains to be done.

The myoglobin fold of mammals is also found in widely separated invertebrates.

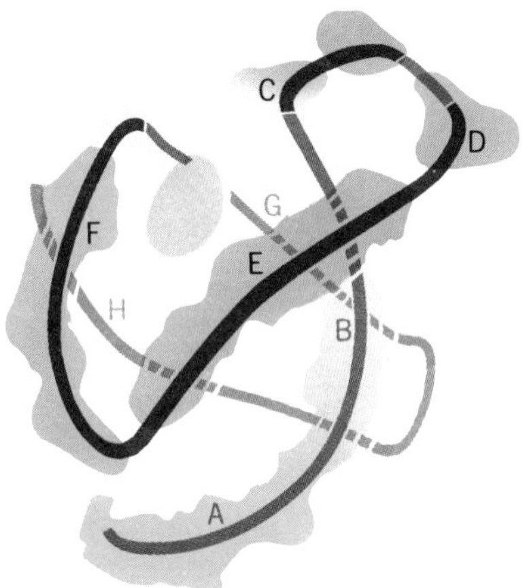

Low-resolution model of an oxygen-binding heme protein from larvae of the fly C. thummi *(Courtesy of R. Huber, Munich.)*

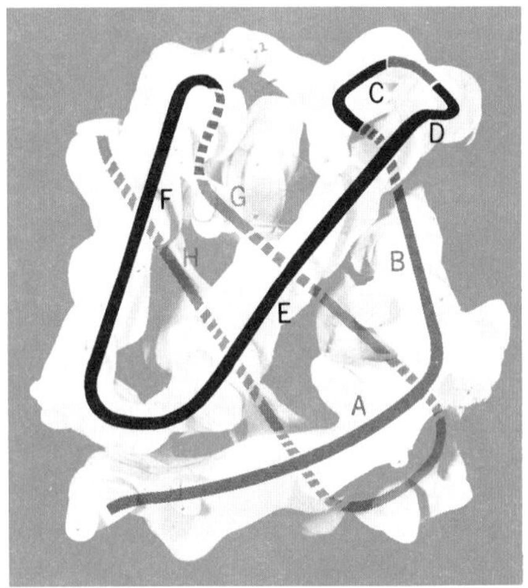

Low-resolution model of a globin from the marine worm G. dibranchiata. *(Courtesy of W. Love, Johns Hopkins.)*

3.5 CYTOCHROME C, THE ELECTRON CARRIER

Cytochrome c is also a carrier molecule like hemoglobin, but instead of an oxygen molecule it carries an electron. It has a heme group like myoglobin, with a polypeptide chain of 104 amino acids wrapped around it and a molecular weight of 12,400. But unlike myoglobin, it is covalently bound to its heme, through two cysteine side chains that form thioether links to what were vinyl side groups on the myoglobin heme. The heme and the section of polypeptide chain to which it is attached are shown to the right. Immediately after the second thioether connection there occurs a His that coordinates to the iron atom much as His F8 does in myoglobin. But the far side of the iron is not open for coordination by small molecules in cytochrome c. Instead, another side chain, whose identity is unknown at present, provides a sixth ligand, so that the iron atom is completely surrounded by an octahedron of ligands. It was once thought that this sixth ligand might be one of the other two His in the chain, or Lys, but chemical evidence from side-chain blocking experiments (35) suggests that it is probably a particular methionine, Met 80.

The function of the iron atom in cytochrome c is not to coordinate a small molecule but to be oxidized and reduced by the appropriate neighboring molecules or complexes. It is part of an energy breakdown and storage system known as the terminal oxidation chain, which is carried out in organelles the size of bacteria, called mitochondria, within the cells. As soon as aerobic organisms evolved, some time between 1 and 2 billion years ago, a molecule that does what cytochrome c does became necessary. It is not surprising, therefore, that cytochrome c is more widely distributed throughout living organisms than is hemoglobin, which did not become necessary until organisms became too big for simple molecular diffusion of oxygen.* What is more surprising is that the cytochrome c molecule has remained effectively unaltered through a billion years of evolution. Cytochrome c from wheat germ or rice sprouts, when extracted and purified, will react with cytochrome oxidase from horses, and horse cytochrome will react with cytochrome oxidase from bakers' yeast. Whatever else has happened to the molecule, the parts essential to its operation as an electron-transfer device have remained the same.

More data are available on the amino acid sequences of cytochrome c of different species than for any other protein, largely as a result of the efforts of Margoliash, Smith, and coworkers. Nearly 35 species have now been sequenced, and more are under way. Most of these sequences are tabulated in the *Atlas of Protein Sequences* (11), and the significance of their comparisons has been well brought out by Margoliash and Smith in a paper in the molecular evolution symposium (28). Over one third of the residues are absolutely identical in every species, from vertebrates of all kinds to insects, yeast, and wheat. The great majority of changes that have occurred are highly conservative, replacing one amino acid by a very similar one. The invariant residues tend to occur in clusters, and one long sequence from residues 70 to 80 is totally unchanging. This sequence must be one of the most vital portions of the molecule, with any alterations in amino acids because of mutations in DNA sequence being immediately lethal. Not surprisingly, the connections to the heme at Cys 14, Cys 17, and His 18 are also invariant.

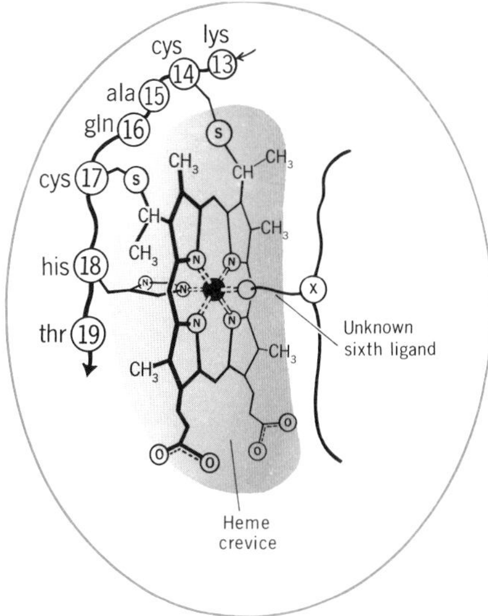

Schematic drawing of cytochrome c. The molecule is egg-shaped, about 25 × 25 × 35 A, with a crevice along one side into which the heme group is fitted.

It is obviously of interest to find out where these changing and invariant regions are on the molecule, as a clue to how the molecule operates. The x-ray structure analysis of horse heart cytochrome c is still at a low-resolution stage, and only the general outlines of the molecule are known (36, 37). The molecule as it appears at low resolution is shown to the right. The chief difference between cytochrome c and myoglobin is that the heme group in cytochrome c sits in a crevice in the molecule, at right angles to the surface, with each side of the heme coordinated to a ligand from a wall of the crevice. The heme is firmly attached to polypeptide chain by means of residues Cys 14, Cys 17, and His 18 on the left, but its connection to the sixth ligand, on the right, seems more

The low-resolution model of cytochrome c on which the schematic is based.

* One of the more intriguing unproved hypotheses about cellular evolution is the possibility that mitochondria might be the gutted remnants of bacteria that once lived in symbiotic partnership with nuclear cells. The mitochondria divide when the cell divides in mitosis rather than being formed anew from nuclear DNA. They have a limited amount of their own DNA and ribosomes that resemble bacterial ribosomes more than those in the cytoplasm of the cell. Moreover, the mitochondrial wall resembles that of bacteria. For more on mitochondria, see Lehninger.

tenuous. The sixth ligand reaches to the heme from what appears to be a loop of extended polypeptide chain.

There appears to be little or no α helix in cytochrome c, and its construction principles are quite different from those of hemoglobin and myoglobin. The molecule appears to be a cage or cocoon of extended chain wrapped around the heme group rather than a framework built up from eight rigid units as in myoglobin. In several places the extended chain bends back and runs parallel to itself in a manner that will be encountered again with the enzymes discussed in Chapter 4. It is possible to make a tentative fitting between the side chains that branch off from these extended stretches and parts of the amino acid sequence. If this is done in the most convincing way, the loop of chain mentioned earlier brings Met 80 past the heme at the sixth coordination site. This, plus the chemical evidence (35) and the total invariance of residues 70–80, makes a very suggestive case for the sixth ligand being Met 80.

The molecule as described above is in its oxidized form. There are many indirect lines of evidence (38) which suggest that there may be a certain amount of change in the folding of the polypeptide chain as the heme is oxidized or reduced, even if only in the immediate vicinity of the heme. The only way to settle this question is to do the x-ray structure analysis of both forms, and this is under way. If it is true, it would be another example, like hemoglobin, of these proteins functioning as machines on a molecular scale, with atoms as the moving parts.

If the differences between sequences of different species are tabulated, the same sort of pattern appears for cytochrome c as for the hemoglobins but without the complication of multiple chain types. The cytochrome c comparison is shown in the table across the page. The primates differ from other mammals by 8–12 residues of 104, and the other orders of mammals differ among themselves by about 5 residues. Mammals and birds differ by an average of 9.9, mammals and reptiles and amphibia by 14, all land vertebrates differ from tuna fish by an average of 18.5, the vertebrates of all kinds differ from insects by 26, and all animals differ from plants and microorganisms by an average of 47 residues. These sequence differences and similarities can be used to construct a family tree of living organisms that is virtually identical to that produced by traditional methods. Such a tree for the species of the table is shown on the opposite page. Although the tree shown is only a rough sketch, very precise ones can be drawn up (30, 39) that give a quantitative picture of the degree of relatedness or separateness of species in a way that is difficult using the more conventional traits of macroscopic taxonomy. The advantage of using protein-sequence comparisons is that a great many such trees can be built up using a great many proteins and cross-checked until their eventual correctness cannot be doubted.

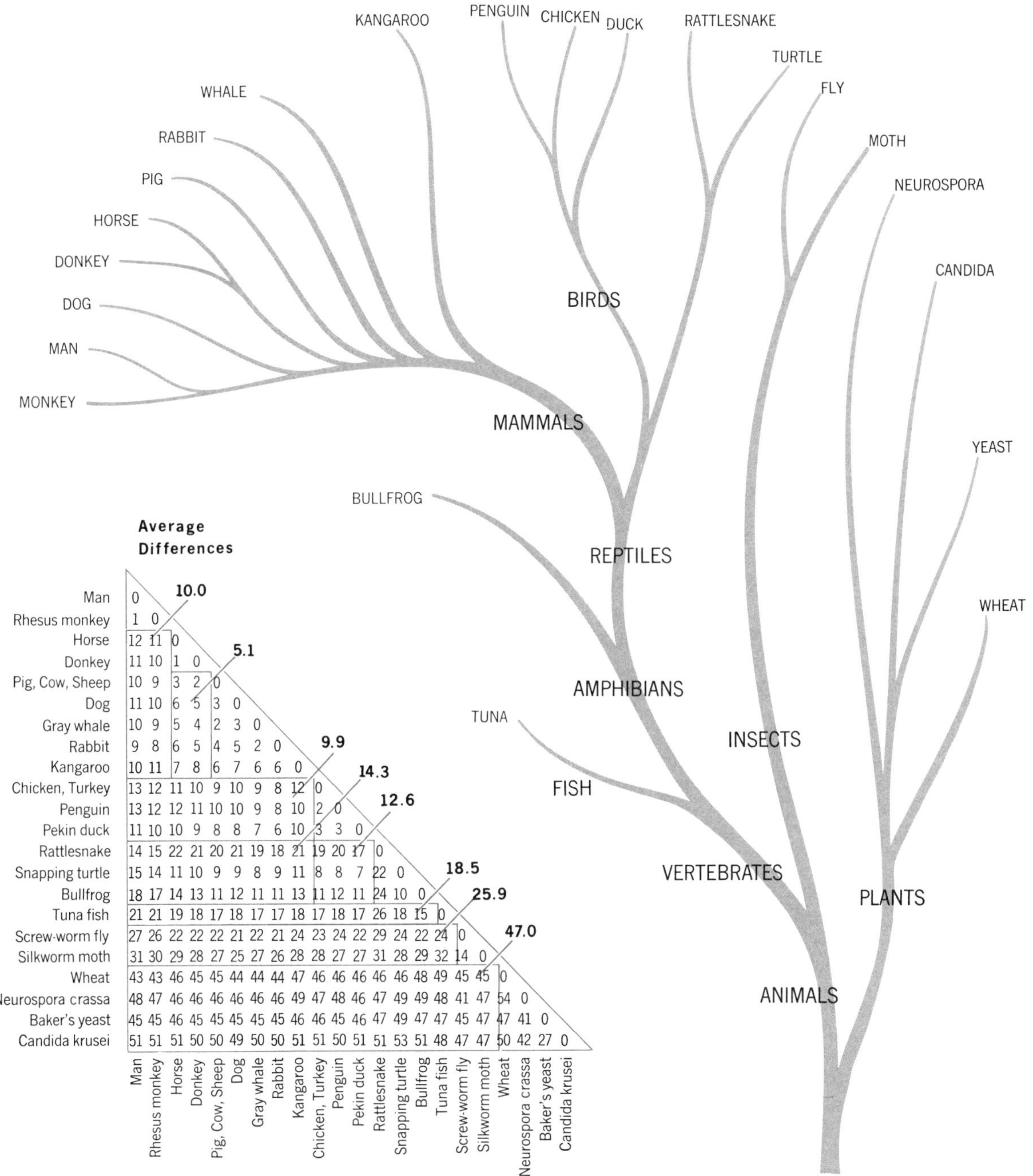

Average Differences

	Man	Rhesus monkey	Horse	Donkey	Pig, Cow, Sheep	Dog	Gray whale	Rabbit	Kangaroo	Chicken, Turkey	Penguin	Pekin duck	Rattlesnake	Snapping turtle	Bullfrog	Tuna fish	Screw-worm fly	Silkworm moth	Wheat	Neurospora crassa	Baker's yeast	Candida krusei
Man	0																					
Rhesus monkey	1	0																				
Horse	12	11	0																			
Donkey	11	10	1	0																		
Pig, Cow, Sheep	10	9	3	2	0																	
Dog	11	10	6	5	3	0																
Gray whale	10	9	5	4	2	3	0															
Rabbit	9	8	6	5	4	5	2	0														
Kangaroo	10	11	7	8	6	7	6	6	0													
Chicken, Turkey	13	12	11	10	9	10	9	8	12	0												
Penguin	13	12	12	11	10	10	9	8	10	2	0											
Pekin duck	11	10	10	9	8	8	7	6	10	3	3	0										
Rattlesnake	14	15	22	21	20	21	19	18	21	19	20	17	0									
Snapping turtle	15	14	11	10	9	9	8	9	11	8	8	7	22	0								
Bullfrog	18	17	14	13	11	12	11	11	13	11	12	11	24	10	0							
Tuna fish	21	21	19	18	17	18	17	17	18	17	18	17	26	18	15	0						
Screw-worm fly	27	26	22	22	22	21	22	21	24	23	24	22	29	24	22	24	0					
Silkworm moth	31	30	29	28	27	25	27	26	28	28	27	27	31	28	29	32	14	0				
Wheat	43	43	46	45	45	44	44	44	47	46	46	46	46	46	48	49	45	45	0			
Neurospora crassa	48	47	46	46	46	46	46	46	49	47	48	46	47	49	49	48	41	47	54	0		
Baker's yeast	45	45	46	45	45	45	45	45	46	46	45	46	47	49	47	47	45	47	47	41	0	
Candida krusei	51	51	51	50	50	49	50	50	51	51	50	51	51	53	51	48	47	47	50	42	27	0

THE FAMILY TREE OF THE CYTOCHROMES C

The species differences shown in the table above left lead to a tree of family relatedness. Note that these is no ascending hierarchy. From the viewpoint of a yeast (if it had one, and therein lies a real if anthropocentric distinction), a moth, a man, and a bullfrog are equally far away. Note also how provincial is the view that we usually take of the living kingdom. The differences between fungi are greater than those between insects and vertebrates.

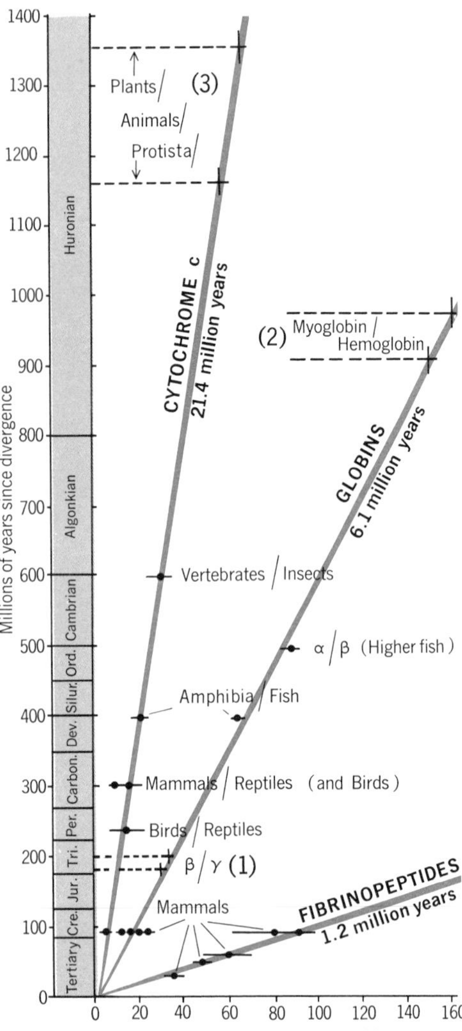

Average rates of change in sequences for three proteins of varying function. Unit evolutionary period *in millions of years is given for each protein. Three critical branch points in evolution (1), (2), (3) can be roughly dated from sequence differences. Repeated changes at the same amino-acid site are accounted for by the expression (32):*

$$\frac{N_{corr}}{100} = -\ln\left[1 - \frac{N_{raw}}{100}\right]$$

Standard names of geological periods are abbreviated at left.

3.6 THE CLOCKS OF EVOLUTION

If more of the available species data on the globins than are given on page 52 are compared with the cytochrome data, it is obvious that the globins are changing more rapidly. Moreover, if the differences between branches of the tree of living organisms for a given protein are plotted against the time in the past when the two branches diverged, a reasonably good straight line is found, with a slope that is characteristic of the protein. Such a plot for cytochrome c, the globins, and the fibrinopeptides appears at the left. Each protein appears to have a characteristic *unit evolutionary period*, or the time required over a long term average for one acceptable amino acid change to appear per 100 residues.

This is not to imply that the genes for these three proteins are mutating at different rates but only that the specifications for working molecules are different in the three cases, and that random changes in sequence have differing probabilities of being lethal. Cytochrome c is a finely tooled molecule that must mesh with other macromolecules as it operates in the mitochondrion, and the vast majority of chance alterations are intolerable. The globins are less sensitive, interacting as they do mainly with smaller molecules, and a chance mutation that is tolerable will occur more frequently. The fibrinopeptides are 20-residue fragments that are chopped out of fibrinogen when it is changed to fibrin during the blood-clotting process. They are analogous to the glass in a fire-alarm box and are scrapped after the clotting mechanism has been triggered. Apparently the restrictions on amino acid residues in these peptides are few, and a large proportion of the mutations are accepted, leading to a rapidly evolving protein. A strategy for studying molecular evolution arises from all this. If the broad outlines of the history of life are wanted, then a widely distributed, slowly evolving protein such as cytochrome c must be studied. On the other hand, rapidly changing proteins such as the fibrinopeptides are most useful in studying the details of one branch of the tree.

The term "clocks of evolution" is perhaps badly chosen, for it suggests some sort of internal mechanism ticking away and unrolling evolution like a script. This, of course, is nonsense. The rate of evolution and of change in protein sequence will change from one period to another as the selective forces on the species change. It is only over such long time spans as those illustrated at the left that this linear approximation has any validity. Nevertheless, because of this average linearity, it may be possible some day to assign precise dates (as geological time figures go) for such distant moments as the separation of plants and animals, the development of photosynthesis and aerobic metabolism, the origin of nuclear cells, and even earlier branch points in the history of life.

CHAPTER FOUR
MOLECULAR CATALYSTS

4.1 WHAT IS AN ENZYME?

In 1890, in a paper on the kinetics of the enzyme invertase, O'Sullivan and Tompson summed up (1):

The group of bodies known as unorganized ferments or enzymes are, as far as the products of their action are concerned, well defined; but, notwithstanding much useful and valuable work, we know very little of their mode of action and far less of their chemical constitution. They are a highly interesting and important class of substances, and are named unorganized ferments in contradistinction to the living or organized ferments, because they possess a life function without life . . . They may be properly called *the transforming* ferments, for their function is a transforming or breaking down of one substance into two or more others. They may be looked upon as the reagents of life, and they are all products of it. Their function is a life function. Is there anything in this which can be distinguished from ordinary chemical action? If so, what?

It was gradually realized that the answer to this was "No." These ferments followed the rules of chemistry and had no special vital properties. Moreover, the "organized ferments" were found to be yeast cultures that operated only because they were full of enzymes themselves. Enzymes were shown to be biological catalysts, which, like any other catalysts, act to change the *rate* at which reactions take place without causing anything to happen that would not occur in time without them. Because they were organic but were too large to pass through the pores of common membranes and parchment filters, they were assumed to be colloids of some sort. It was also recognized that part of enzyme action involved binding or adsorbing the substrates—the molecules on which the enzyme acted—to the enzyme surface, in agreement with the colloid theory. The British biochemist Bayliss stated quite confidently in 1914 that enzymes were *not* proteins (2). Even as late as 1930, the very respected biochemist J. B. S. Haldane had to admit (3): "The attempts to purify enzymes have led to no definite conclusion as to their chemical nature. Except Sumner's urease, none of the most highly purified preparations appeared to be proteins, though all so far analysed contain C, H, O, and N." Yet within four years the first x-ray diffraction photographs (4) were taken of a pure, crystalline protein enzyme, and the long process of deciphering the molecular structures of enzymes began.

Every enzyme now known has turned out to be a globular protein, although many require other metal ions and organic molecules, or cofactors, to operate. Like all catalysts, they speed the approach to chemical equilibrium without shifting the equilibrium point itself. They do this by providing alternative pathways for the reaction—pathways that have a lower energy of activation barrier and hence are traversed more rapidly. Enzymes have an important control function; if several possible reactions use the same starting material and if only one of these is accelerated by catalysis, then obviously most of the starting material will follow this path. Although enzymes cannot cause anything to occur that would not occur otherwise, they can make one reaction occur at the expense of another and hence can play a switching role.

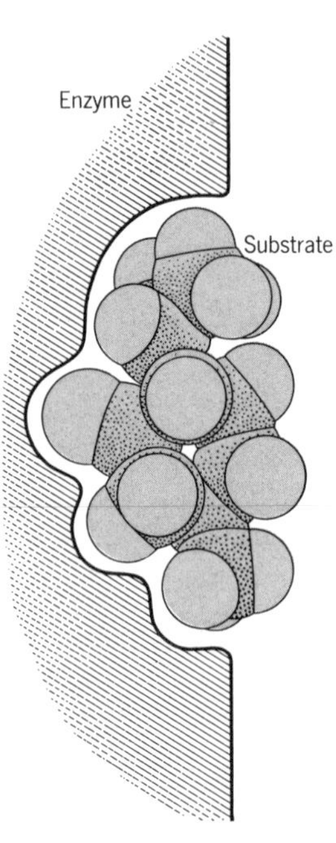

Emil Fischer proposed a "lock and key" fitting of substrate to enzyme, in which the shape of the surface of the enzyme was modeled to that of the molecules that bind to it. Such a picture could not be verified before the days of x-ray analysis of protein structure. But now a more subtle form of this fitting has been found in lysozyme and carboxypeptidase, and is expected in other enzymes.

Enzymes are usually quite specific in their action. A very small change in a side group in a molecule may mean that the enzyme will no longer accept it as a substrate. This led Emil Fischer in 1894 to propose his famous "lock and key" metaphor (5), which says that the surface of an enzyme is shaped in such a way that its substrate fits onto it like a key into a lock, and that only slightly altered molecules may be unable to fit onto the enzyme and hence will not be acted upon by it. It was once thought that this binding might be the whole story of the speedup of reactions by enzymes; two reactants which otherwise would only have a certain probability of reaction on collision in solution would be bound in close proximity on the surface of the enzyme, where their chances of reacting would be greatly enhanced. But it was soon realized that this "proximity factor," by itself, failed by many orders of magnitude in explaining enzymic reaction rates. There are two separate problems involved in the mode of action of enzymes: the great specificity or selectivity of an enzyme in binding substrates, and the mechanism of catalytic action itself. It is easy to imagine molecules which contain bonds that the enzyme would break if only the molecule would bind properly, and it is not difficult to find other molecules that will bind to the active site of an enzyme but will not then react. The latter molecules are known as competitive inhibitors, as they compete with the true substrate for the enzyme.

The specificity problem looks like a natural one for attack by x-ray structure analysis, for if any "lock and key" accommodations of enzyme and substrate exist, they should be visible. It is not as certain that a knowledge of the structure of an enzyme, even with its substrate in place, would necessarily tell much about the actual mechanism of catalytic action. Yet Phillips and his associates, who solved the first enzyme structure, have been quite successful in proposing a catalytic mechanism for lysozyme which so far is consistent with all the chemical work that preexisted or has been stimulated by x-ray analysis.

4.2 LYSOZYME

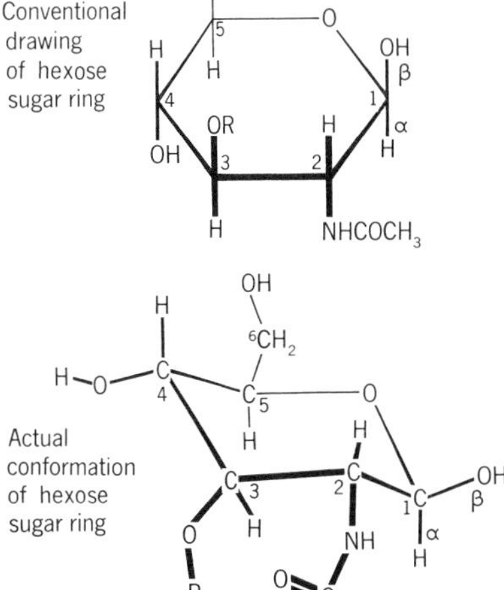

In 1922 Alexander Fleming persisted in working in his laboratory even though suffering from a cold, and, in a somewhat offhand manner, added a few drops of nasal mucus to a bacterial culture plate to see what would happen. To his great surprise, he found a few days later that something in the mucus was killing the bacteria. This substance was the enzyme lysozyme.* It has since been found in most bodily secretions and in great quantities in the whites of eggs. Hen egg white enzyme is a globular protein of molecular weight 14,600, with 129 amino acids in a single polypeptide chain and four disulfide bridges.

Lysozyme attacks many bacteria by lysing, or dissolving, the mucopolysaccharide structure of the cell wall. The basic unit of the polysaccharide is shown at the right. This is a hexose sugar ring, which can be polymerized by linking its C_1 and C_4 carbons in two ways. If the C_1 hydroxyl group that links to the C_4 of the next ring is in the α position, the polymer is called an α-1,4-polysaccharide, of which starch is an example. In contrast, cellulose, chitin, and the bacterial polysaccharides are built from β-hydroxyl monomers as shown here and are called β-1,4-polymers. The bacterial cell wall polymer is an alternating polymer of NAG, N-acetylglucosamine (with a hydroxyl group at —OR), and NAM, N-acetylmuramic acid [with —OCH(CH_3)COOH for —OR] (lower right). These chains are cross-linked by short polypeptides that are attached to the —OR side chains of NAM by peptide bonds. Lysozyme cuts the polysaccharide chain in a specific place, on the far side of a linking oxygen atom that is attached to the C_4 of NAG. For an alternating -NAG-NAM-NAG-NAM- polymer, the cut is between a C_1 of NAM and the chain-linking oxygen; for the artificial substrate poly-NAG, any bond between a C_1 atom and its linking oxygen can be cut. Poly-NAM is not a substrate, for reasons that will become clear from the molecular structure.

Lysozyme was the second protein and the first enzyme to have its detailed molecular structure worked out by x-ray analysis, by Phillips, North, Blake, and coworkers (10–16). The NAG-NAG-NAG trimer was found to form a stable inhibited complex, and the structure of the enzyme was worked out with and without the tri-NAG present. Longer NAG polymers were substrates, being cleaved with greater and greater rapidity up to the hexasaccharide. Although it was not possible to observe the enzyme with more than the trimer in place, it was possible by model building to work out

—OR = —OH in NAG
—OR = —OCH(CH_3)COOH in NAM
—OR = —OCH(CH_3)—CONH—CHR_1—CONH—CHR_2—CONH—CHR_3— . . . in bacterial cell walls. R_1, R_2, R_3 are amino acid side groups.

The substrate of lysozyme is an alternating copolymer of NAG and NAM.

* Penicillin, Fleming's other great discovery seven years later, was found by accident because Fleming was delinquent in washing up old Petri dishes from unsuccessful experiments. The moral with regard to the systematic nature of research need not be spelled out.

the probable mode of binding of the longer polysaccharides, and from this to suggest a catalytic mechanism.

The amino acid sequences of hen egg white and human lysozyme and bovine α-lactalbumin (a milk protein) are compared below. As we shall see later, the lysozymes and α-lactalbumin appear to be evolutionary second cousins, probably with very much the same polypeptide chain folding. The folding of the polypeptide chain in lysozyme is shown on the opposite page and in the stereo drawing on page 72. The most striking feature of the molecule is the crevice running across the waist of the egg-shaped molecule. It divides the molecule into two parts, a predominately helical core containing residues 1–40 and 101–129, and a more irregular wing from 41 to 87, with a helix from 88 to 100 binding the two halves.

The structure of the enzyme with its bound tri-NAG inhibitor is on page 72. It is obvious that lysozyme is a different sort of molecule than myoglobin. There is much less helix. Three fairly regular α helices do exist, made up of residues 5–15, 24–34, and 88–96. Helix 5–15 is a reasonable Pauling–Corey α helix. But 24–

	1	2	3	4	5	6	7	8	9	10	11	12	13	14	15	16	17	18	19	20	21	22	23	24	25	26	27
Chicken	lys	val	phe	gly	arg	cys	glu	leu	ala	ala	ala	met	lys	arg	his	gly	leu	asp	asn	tyr	arg	gly	tyr	ser	leu	gly	asn
Human	lys	val	phe	glu	arg	cys	glu	leu	ala	arg	thr	leu	lys	arg	leu	gly	met	asx	gly	tyr	arg	gly	ilu	ser	leu	ala	asx
α-Lact.	glu	gln	leu	thr	lys	cys	glu	val	phe	arg	glu	leu	lys	......	......	asp	leu	lys	gly	tyr	gly	gly	val	ser	leu	pro	glu
Cys link						127																					

	28	29	30	31	32	33	34	35		36	37	38	39	40	41	42	43	44	45	46	47		48	49	50	51	52
Chicken	trp	val	cys	ala	ala	lys	phe	glu	......	ser	asn	phe	asn	thr	gln	ala	thr	asn	arg	asn	thr	......	asp	gly	ser	thr	asp
Human	trp	met	cys	leu	ala	lys	trp	glu	......	ser	gly	tyr	asn	thr	arg	ala	thr	asx	tyr	asx	ala	gly	asx	arg	ser	thr	asp
α-Lact.	trp	val	cys	thr	thr	......	phe	his	thr	ser	gly	tyr	asx	thr	glx	ala	ilu	val	glx	asx	......	......	asx	glx	ser	thr	asx
Cys link			115																								

	53	54	55	56	57	58	59	60	61	62	63	64	65	66	67	68	69	70	71	72	73	74	75	76	77	78	79
Chicken	tyr	gly	ilu	leu	gln	ilu	asn	ser	arg	trp	trp	cys	asn	asp	gly	arg	thr	pro	gly	ser	arg	asn	leu	cys	asn	ilu	pro
Human	tyr	gly	ilu	phe	gln	ilu	asx	ser	arg	tyr	trp	cys	asx	asx	gly	lys	thr	pro	gly	ala	val	asn	ala	cys	his	leu	ser
α-Lact.	tyr	gly	leu	phe	glx	ilu	asx	asx	lys	ilu	trp	cys	lys	asx	asx	glx	asx	pro	his	ser	ser	asn	ilu	cys	asn	ilu	ser
Cys link												80												94			

	80	81	82	83	84	85	86	87	88	89	90	91	92	93	94	95	96	97	98	99	100	101	102	103	104	105	106
Chicken	cys	ser	ala	leu	leu	ser	ser	asp	ilu	thr	ala	ser	val	asn	cys	ala	lys	lys	ilu	val	ser	asp	gly	asp	gly	met	asn
Human	cys	ser	ala	leu	leu	glx	asx	asx	ilu	ala	ala	asx	val	ala	cys	ala	lys	arg	val	arg	asx	pro	......	glx	gly	ilu	arg
α-Lact.	cys	asp	lys	phe	leu	asx	asx	asx	leu	thr	asx	asx	ilu	met	cys	val	lys	lys	ilu	leu	asp	lys	......	val	gly	ilu	asn
Cys link	64														76												

	107	108	109	110	111	112	113	114	115	116	117	118	119	120	121	122	123	124	125	126	127		128	129
Chicken	ala	trp	val	ala	trp	arg	asn	arg	cys	lys	gly	thr	asp	val	gln	ala	trp	ilu	arg	gly	cys	......	arg	leu
Human	ala	trp	val	ala	trp	arg	asn	arg	asx	val	arg	gln	tyr	val	glx	......	......	......	......	gly	cys	......	gly	val
α-Lact.	tyr	trp	leu	ala	his	lys	ala	leu	cys	ser	glu	lys	leu	asp	gln	trp	trp	leu	......	......	cys	glu	lys	leu
Cys link									30												6			

CHICKEN EGG-WHITE LYSOZYME SEQUENCE (6, 7). *Human lysozyme (preliminary sequence data) from (8); bovine lactalbumin from (9). Identical residues in two or more sequences are tinted. Dots indicate a gap in a sequence, either from a deletion in one sequence or an insertion in another. Where the state of amidation of a side chain is not yet certain, Glx and Asx are used, and the most favorable choice (Glu or Gln, Asp or Asn) is assumed in drawing homologies. The numbering is that of chicken lysozyme. It is now known (8) that human lysozyme contains eight half-cystines (Cys), including one at position 115.*

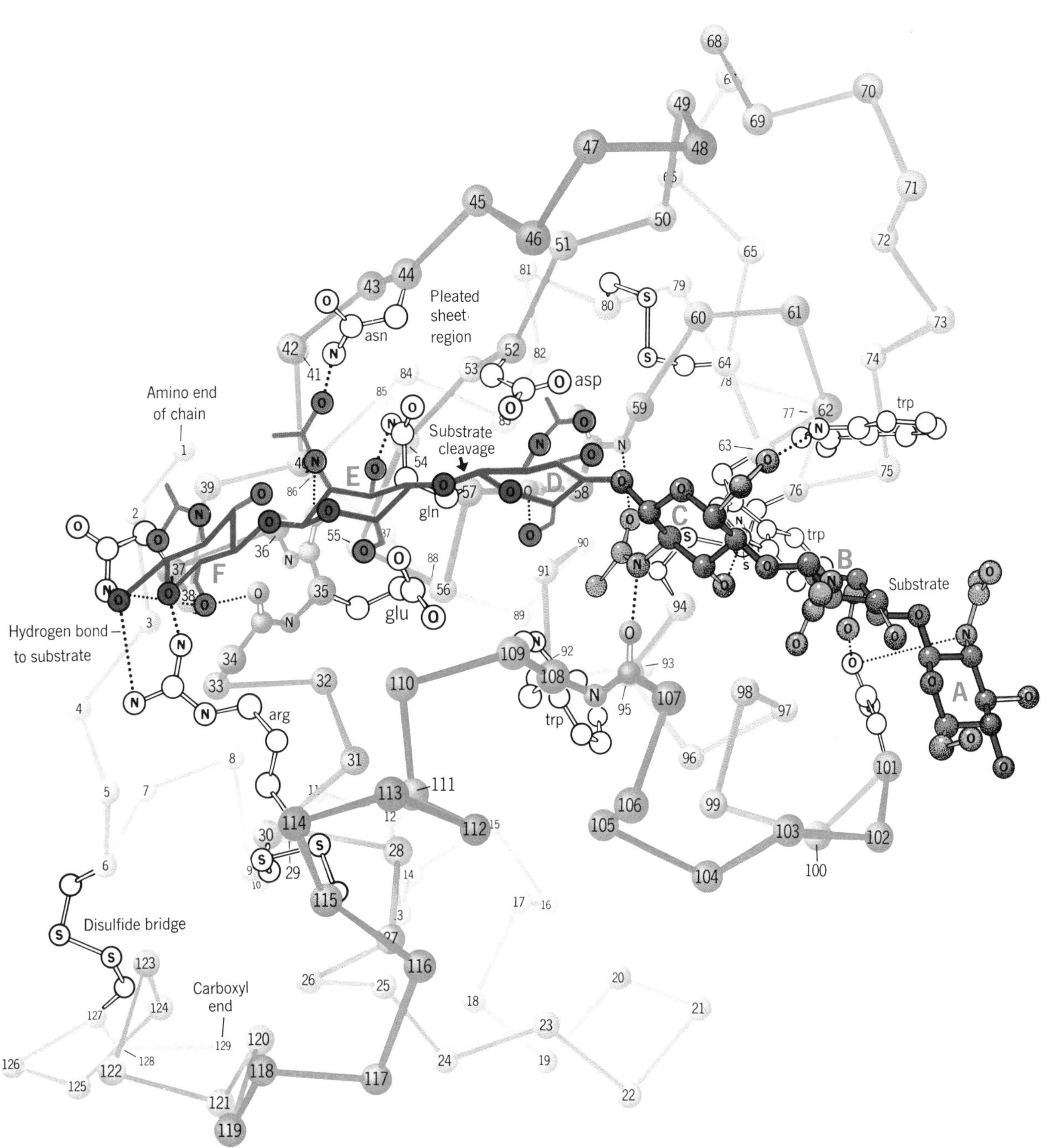

THE FOLDING OF THE MAIN CHAIN IN CHICKEN LYSOZYME. *The crevice that forms the active site runs horizontally across the molecule. The hexasaccharide substrate is shown in a darker color. Rings A, B, and C to the right come from the observed trimer binding. Rings D, E, and F are inferred from model building. The side chains that are believed to interact with the substrate are shown in line. Ilu 98 is so bulky that it helps to prevent NAM, with its large side group, from binding at ring position C, and thus establishes the arrangement on the molecule of the alternating -NAG-NAM- copolymer and points out the locus of cleavage in the active site. (Coordinates by courtesy of Dr. D. C. Phillips, Oxford.)*

34, and to a lesser extent 88–96, have their carbonyl groups rotated out from the helix axis and their NH groups turned in so as to point, not at the carbonyl group four residues back, but between the third and fourth carbonyl back. The hydrogen bonds are therefore intermediate between those of an α helix and a 3_{10} helix (see page 29). Another short helix, 80–85, is even more like a 3_{10} helix in its hydrogen bonding. As in myoglobin, so the last residue of the 5–15 helix is tightened up to the 3_{10} bonding pattern (diagram, page 50). This tightening may turn out to be a common means of ending an α helix and sending the polypeptide chain off in another direction. There are tendencies for the chain to adopt a helix conformation of the same general dimensions and Ramachandran coordinates as the α or 3_{10} helix in other places, notably residues 60–63, the continuation of the α helix at 97–101, and 109–115. This particular configuration region in the Ramachandran plot appears to be a favored one because of side-chain packing, even in the absence of perfect α-helix hydrogen bonding.

The first 40 residues form a compact core built around two lengths of helix. But thereafter the molecule begins to build a new structural feature not found in the globins: a β sheet. As shown to the right, the chain runs straight for residues 41–48 and then doubles back for 49–54 to run antiparallel with itself and form several hydrogen bonds like those in the antiparallel pleated sheet. After burying two hydrophic groups (55 and 56) in the bottom of the crevice, it doubles back once more and forms a much less regular third strand to the sheet, before going off in another direction at 61. This new structural feature in globular proteins was suggested in cytochrome c; has been found in ribonuclease, chymotrypsin, and papain; and appears in its ultimate glory in carboxypeptidase.

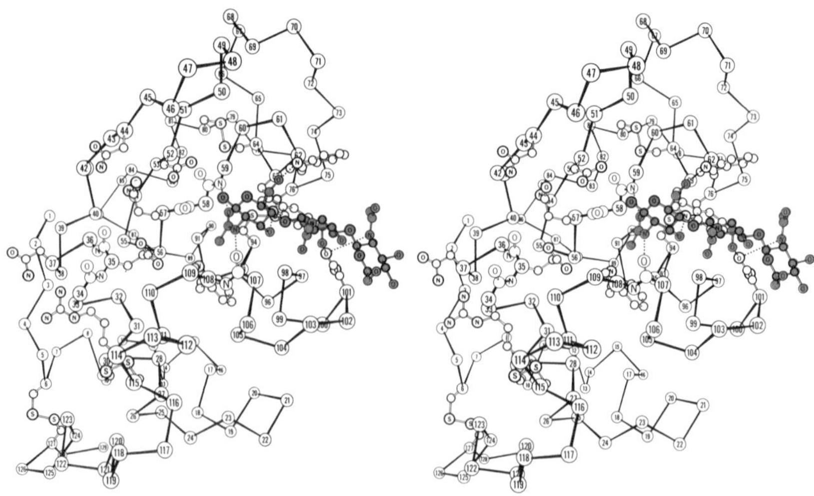

LYSOZYME IN STEREO

Same orientation as on page 71. The trimer inhibitor (in color) and side chains that interact with the full substrate are shown.

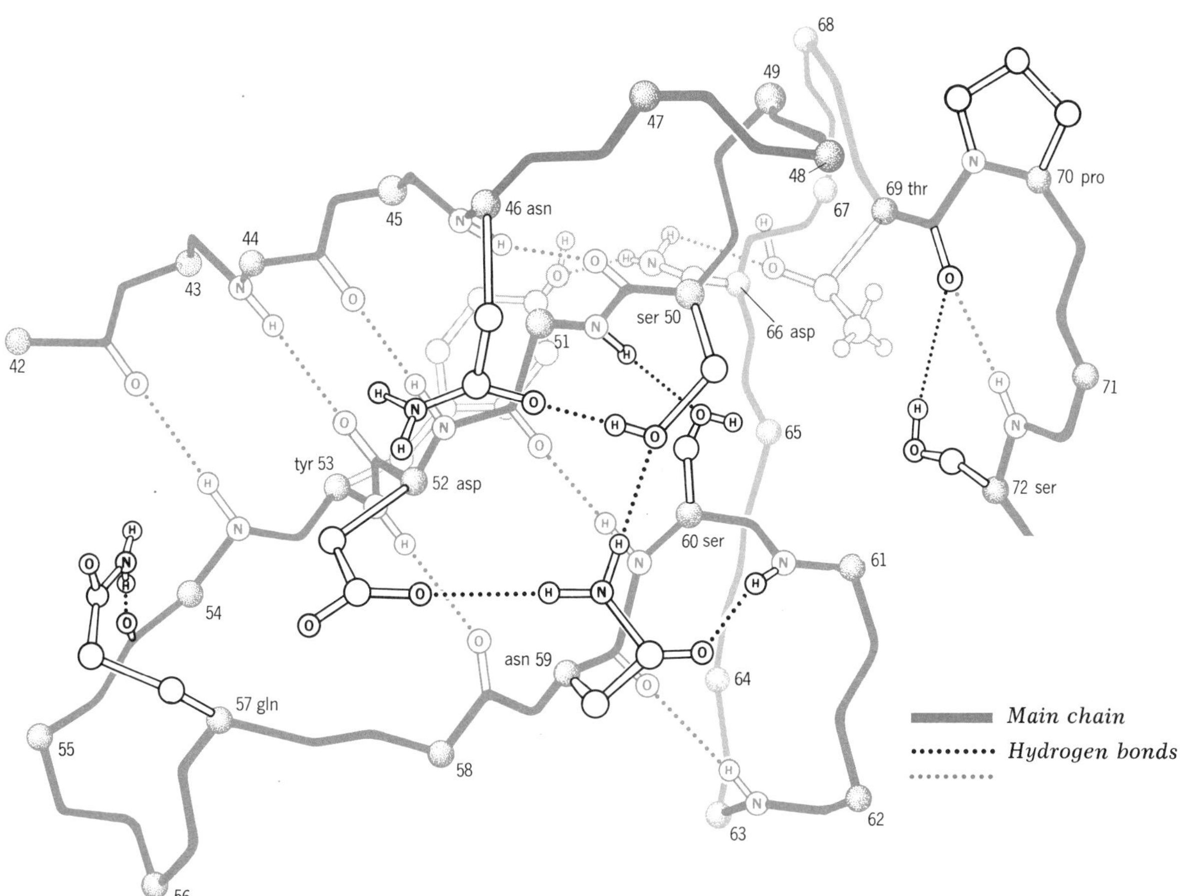

β SHEET STRUCTURE IN LYSOZYME
Main-chain skeleton in color, with α carbons numbered, and with carbonyl oxygens and amide nitrogens shown only when they participate in hydrogen bonding. Note the extensive use of side-chain Ser, Thr, Asn, and Gln for structural bonding, and the way in which Pro 70 forces a bend in the chain.

Lysozyme, like the globins, conforms to the principle of "hydrophobic in, hydrophilic out," or the oil-drop model of a protein. All its charged polar groups are on the surface, as are its uncharged polar groups, with one or two exceptions. The great majority of its nonpolar, hydrophobic groups are buried in the interior. Helices 5–15 and 24–34 have hydrophobic sides that pack against one another in the hydrophobic core formed by the first 40 residues. Helix 88–100 also has a hydrophobic side that is similarly buried in the bottom of the crevice. In contrast, the β sheet is entirely hydrophilic, and this may be one device for directing the proper folding.

How vital this β sheet is to the integrity of the enzyme may be seen by comparing the three sequences from widely separated sources on page 70. Of the 20 residues in the β sheet, 42 to 61, no fewer than 11 are identical in all three proteins, including every side chain that takes part in cross-chain bonding (above). In contrast to this 50 percent retention, only 30 percent of the α-helical residues (11 of 37) and 25 percent of the residues that are in neither α or β regions (17 of 75) are identical in all three chains. Note also that the variations in chain length at residue 47 occur at the exposed end of the hairpin turn, where adjustments would be easier.

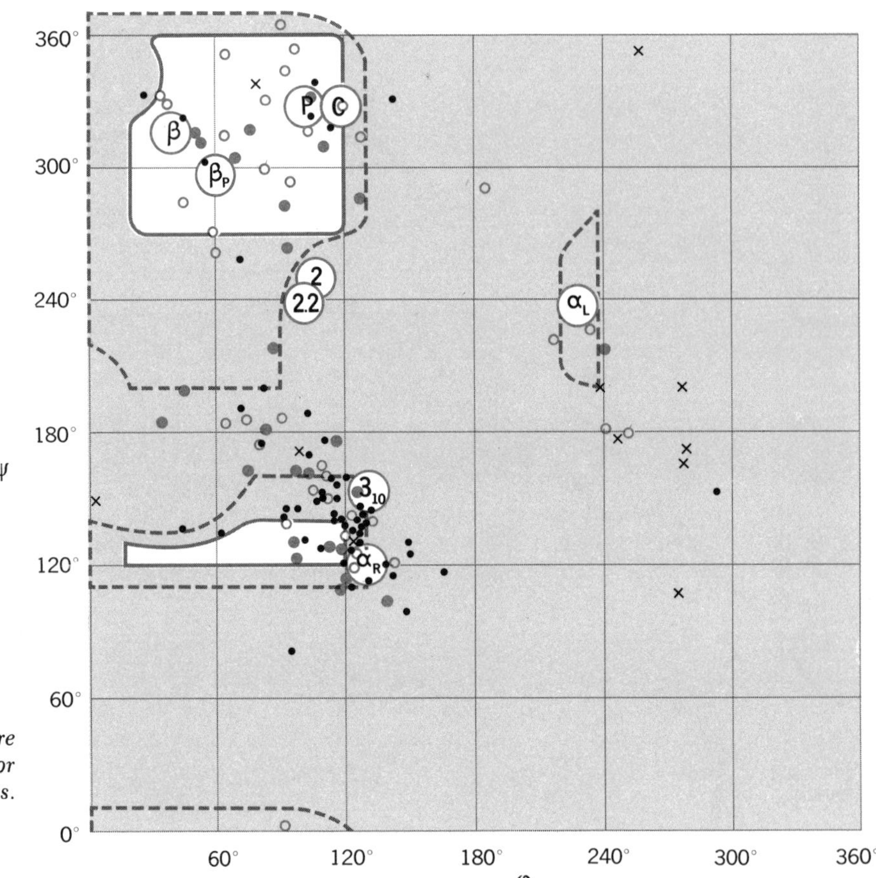

The Ramachandran plot of polypeptide chain conformation in lysozyme. Hydrophobic residues are black dots; hydrophilic, solid color dots; ambivalent, open color dots. Glycine is denoted by crosses.

The Ramachandran plot of lysozyme, above, is strongly reminiscent of myoglobin. Hydrophobic, polar, and charged residues, and Gly, are individually distinguished. As in myoglobin, the zone to the upper left of the α and 3_{10} helices, leading to the β zone, is not strongly forbidden. The zone boundary, such as it is, is nearer ψ=230° than 180°. But the region of carbonyl oxygen clash, from ψ=0 to 100°, is still virtually empty. The deviant residues with φ greater than 180° again are mainly Gly, with no side group. The nonpolar, hydrophobic residues cluster around the helix zone. The β sheet is too open and exposed for them, and in lysozyme their most favored place is on the inner surfaces of helices. The uncharged polar residues are mostly in the β configuration, and the charged groups are evenly distributed in β regions and on the exposed sides of helices.

In myoglobin, all 11 Ser and Thr are hydrogen-bonded to main-chain carbonyl oxygens, but only one of the 7 Asn and Gln is hydrogen-bonded (17). In lysozyme this interchain cross bonding is even more pronounced, and hydrogen bonds involving these residues appear to play an important role in shaping the folding of the chain, especially in the β-sheet region. Eight of the 17 Ser and Thr appear to be involved, and 6 of the 16 Asn and Gln (12). Two Trp and one

Tyr form hydrogen bonds, and Lys 13 forms a salt link to the carboxyl terminal end of the chain. These hydrogen-bond cross-links are markedly clustered. Some occur in the helical core: the bond Ser 24—NH26 starts a helix and Trp 28—CO23 orients it. Bond Asn 27—Trp 111 fastens down the last part of the chain, and CO 115—Thr 118 sends the final tail in the proper direction. A second cluster occurs at the "hinge" between the two halves of the molecule, with Asn 39—NH 41 and with a three-way bond: NH 38 —Ser 36—CO 55. In agreement with their important roles in holding the folded chain together, Ser 24, Trp 28, Asn 39, and Ser 36 are present in all three chains on page 70. But the most striking use of cross-chain hydrogen bonds is in the β sheet wing, from residues 42–72. These are diagrammed on page 73. It looks as if the role of Ser and Thr, Asn and Gln, is to define the folding of this part of the chain more exactly than the β-sheet bonds could do alone. The combination of polar character and hydrogen-bonding ability may make these residues the determinants of exposed β-sheet folding in globular proteins.

The fact that the tri-NAG inhibitor binds in the right half of the crevice as the molecule is shown on page 72 suggests that the crevice contains the active site. Most of the hydrophobic groups that are on the outside of the molecule appear as lining for the crevice, making it a plausible binding site for an organic molecule. There are small changes in side-chain position when the trimer inhibitor (and by inference, the substrate) binds, although nothing like the motion of subunits in hemoglobin. Trp 62 moves about 0.75 Å toward the trimer B ring, and the entire side of the crevice closes down slightly on the substrate. It is possible to build three more

A globular protein molecule is not an open framework, as the main-chain folding drawings might suggest, but is a tightly packed solid. The models below are seen from the right side of the stereo on page 72. Note the closeness of fit of the hexasaccharide substrate, in good Fischer lock-and-key fashion. Some of the essential side chains for substrate binding and catalysis are marked in the left drawing below: Asp 52, Trp 62, Trp 63, and Asp 101. (Model photographed by courtesy of Dr. J. A. Rupley, University of Arizona.)

Space-filling CPK model of lysozyme. Left: enzyme without substrate, showing active site crevice. Right: enzyme-substrate complex, with substrate in color.

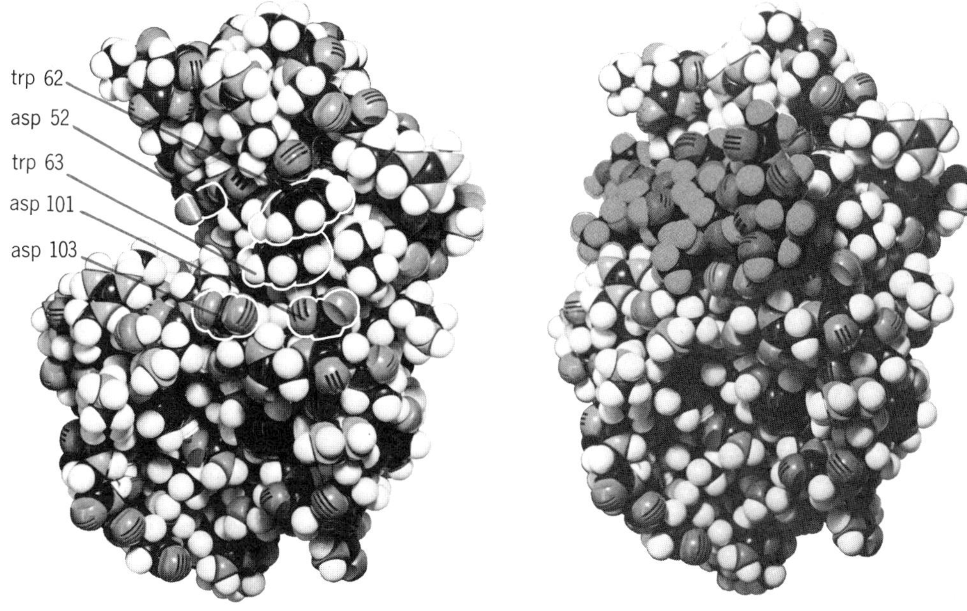

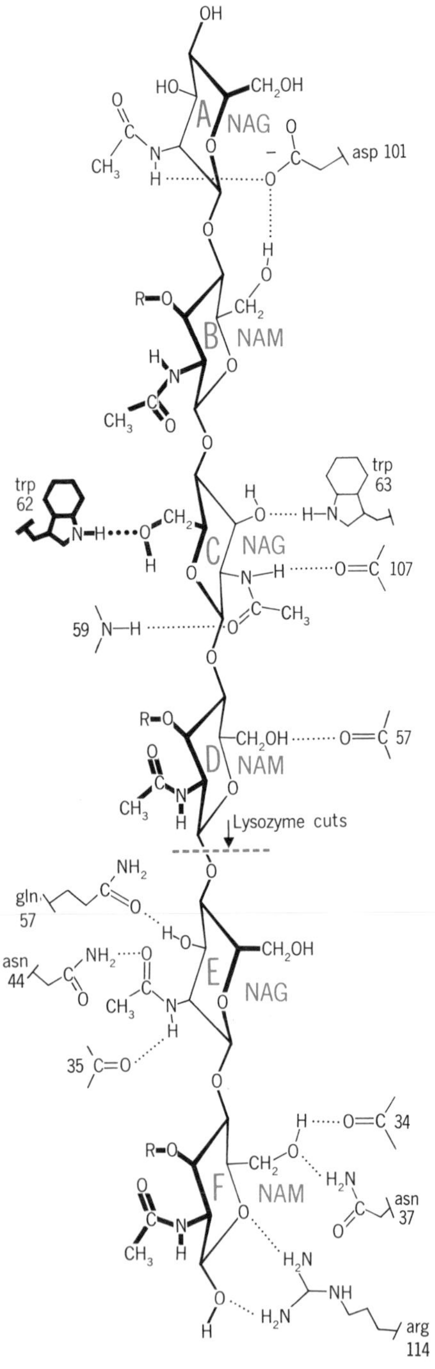

Substrate interactions with lysozyme. The view is into the active crevice, with the darker edges of the rings exposed to the outside, the lighter ones buried at crevice bottom.—OR = —O—CH(CH_3)—COOH.

sugar rings into the left half of the crevice in a convincing manner to form a hypothetical binding for the hexasaccharide substrate. The side groups or main-chain atoms that interact with the trimer or are assumed from model building to interact with the hexamer are shown on pages 71 and 72. A schematic view of the hexamer and its interactions with the enzyme appears on the left. The view is nearly straight into the crevice, with the heavy edges of the sugar rings on the exterior. Although the trimer itself was tri-NAG, which lacks the bulky side chains of NAM at C_3, it was possible to tell which of the sugar-binding sites on the enzyme were NAG and which were NAM, and to identify the region on the enzyme where catalysis must take place. Rings A, C, and E can only be NAG when the alternating copolymer is bound, for these rings extend their C_3 side chains toward the bottom of the crevice. In ring C, in particular, there is no room for the bulky —OCH(CH_3)COOH group of NAM, which would clash with Ilu 98 (see page 71). In rings B, D, and F, these bulky groups extend into the solvent. Since the chain is cut just above a NAG, as drawn here, the catalytic point must be between rings B and C, or D and E. But the first possibility is ruled out because the trimer is not split, and the site of catalysis must be between D and E.

When a search is made in the vicinity of the D/E ring link for possible catalytic groups, two candidates come to light: Asp 52, in a polar environment where it will be ionized, and Glu 35, in largely hydrophobic surroundings where it might easily remain un-ionized. Ring D is not perfectly happy where it is; carbon C_6 and its hydroxyl group are too close to the main-chain carbonyl of residue 52, to Trp 108, and to the —NHCOCH_3 side chain of ring C. The strain can be relieved by warping ring D from a chair conformation to the "half-chair" shown in (b) to the right. This moves the oxygen atom away from the enzyme and swings the —CH_2OH group to an axial position instead of an equatorial one. It can then hydrogen-bond either to the carbonyl of 57 or to Glu 35.

The suggested sequence of events in the catalytic breaking of the C_1—O bond is shown on the opposite page. The un-ionized proton of Glu 35 first attacks the linking oxygen and weakens the C_1—O bond (a). If the bond were to break, ring D would form a carbonium ion. Lemieux and Huber have suggested that under such circumstances in a hexasaccharide the carbonium atom C_1 would share its charge with the ring oxygen, and the ring would adopt a half-chair configuration with carbons C_2, C_1, C_5, and the ring O in a plane (18). The enzyme itself favors this conformation, and the nearby charged Asp 52 helps to stabilize the carbonium ion. These factors contributed by the enzyme lower the activation barrier for the bond-breaking process (b). The Glu 35 proton is replaced by another from an ionizing water molecule, and the resultant hydroxyl ion then attacks the carbonium ion and completes the reaction (c).

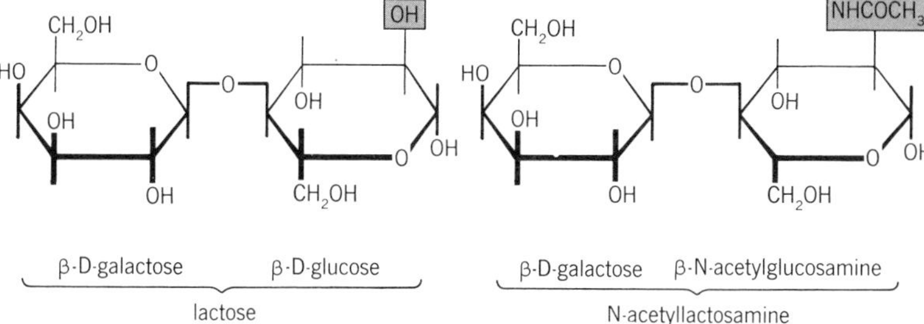

In the absence of α-lactalbumin, the A protein makes N-acetyllactosamine; in its presence, it makes lactose. Differences between products indicated in color.

This proposal of Phillips and coworkers, far from supplanting enzyme chemistry, has suggested a great number of chemical and kinetic experiments, which are being carried out in several laboratories. So far the results of the chemical work have agreed with this mechanism. Rupley, for example, has shown that the NAG hexamer is cleaved between the fourth and fifth ring (19), and Raftery has provided strong evidence for the correctness of the carbonium ion mechanism (20). Koshland (20a) and Raftery (20b) have each shown that Asp 52 is essential for catalysis.

This section cannot end without mention of an incomplete story with fascinating evolutionary possibilities. When Hill and coworkers, following a suggestion of Brew and Campbell (21), solved the amino acid sequence of α-lactalbumin (9), a milk protein of previously unknown function, they found that 53 of its 124 residues were identical to those in hen egg white lysozyme, the four disulfide bridges occurred in the same places, and many more residues could be classed as conservative substitutions. The sequences are compared on page 70, and it is obvious that some sort of family relationship is involved. Phillips, North, and Brown have been able to build the α-lactalbumin sequence quite well into the lysozyme folding pattern, and an independent structure analysis of α-lactalbumin is now underway (22).

α-Lactalbumin by itself is not an enzyme but was found to be one component of a two-protein lactose synthetase system, present only in mammary glands during lactation (23). The other component (the "A" protein) had been discovered in the liver and other organs as an enzyme for the synthesis of N-acetyllactosamine from galactose and NAG. But the combination of the A protein and α-lactalbumin synthesizes the milk sugar lactose from galactose and glucose instead. The noncatalytic α-lactalbumin evidently acts as a control device to switch its partner from one potential synthesis to another.

Lactose is found only in milk and a few plant tissues, although the A protein carries out its other synthesis in many places where lactose is not made. Furthermore, as pregnancy progresses, the

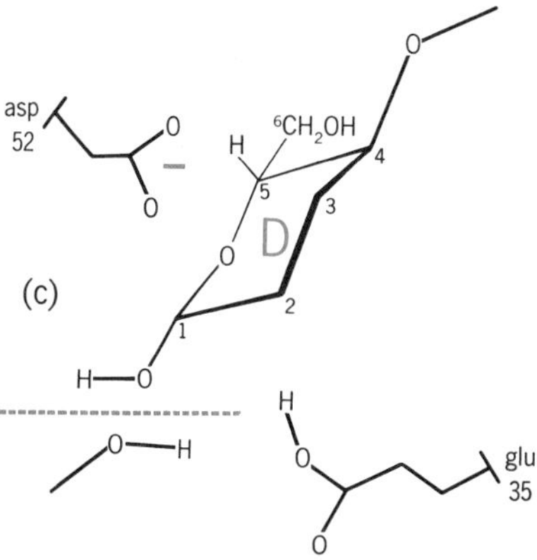

The mechanism of cleavage of the polysaccharide bond in lysozyme.

If α-lactalbumin diverged from lysozyme during the rise of the mammals, then its evolutionary tree would be as at the left; yet the observed differences in sequence on page 70 require the tree at the center if a constant rate of sequence change is assumed. *This assumption is invalid during the evolution of a new protein, and the probable explanation, involving rapid evolution of α-lactalbumin, is at the right.*

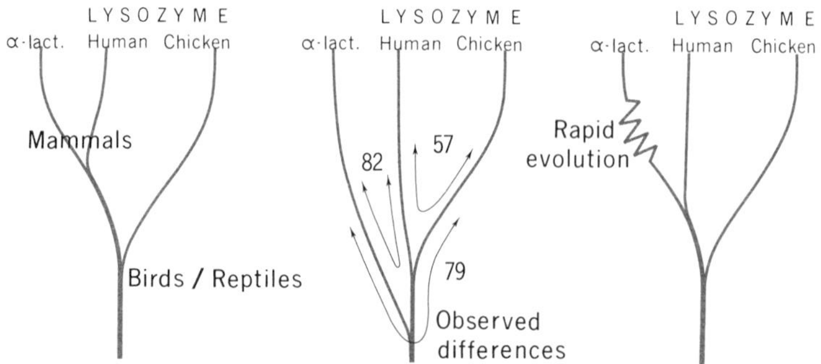

amount of A protein in the mammary glands increases at a steady rate, without milk production. It is only when specific hormones trigger the rapid synthesis of α-lactalbumin late in the process that lactose is synthesized and milk production begins.

What is the relationship between lysozyme and α-lactalbumin? Only the outlines of an answer can be given now, but it appears that when a milk-producing system was being developed during the evolution of the mammals, and when a need for a polysaccharide-synthesizing enzyme arose, a suitable one was found in part by modifying a preexisting polysaccharide-cutting enzyme. This adaptation of preexisting structures to new uses has been seen many times in macroscopic anatomy, as when fins become feet and ultimately wings and hands, or when reptilian jaw hinge bones become the anvil and stirrup of the mammalian inner ear. But this is one of the first documented examples of the same process at the molecular level.

A simple-minded application of the "clocks" ideas of Chapter 3 to these lysozymes and α-lactalbumin leads to an apparent contradiction (above). If α-lactalbumin evolved from a mammalian lysozyme during the course of the development of mammals, then it and human lysozyme should be more similar than either is to hen lysozyme. Conversely, the assumption that rates of change have been constant in all three proteins since divergence leads to the conclusion that the α-lactalbumins separated from the lysozymes long before the first appearance of terrestial vertebrates. Where is the fallacy?

The fallacy, of course, is in the assumption of unchanging rates of accumulation of tolerable mutations. For one particular protein, performing much the same task in a wide spectrum of species, this may be a valid working hypothesis. But when circumstances arise in the environment such that a duplicated gene is being altered, the better to perform a *new* function, selection pressure is unusually severe and changes in sequence will be unusually rapid. This is not only true at the molecular level; one of the frustrations of the paleontologist has always been that of trying to find the relatively scarce transitional varieties between two species, each well adapted and prolific in its own surroundings.

4.3 OTHER CATENASES: RIBONUCLEASE

No other enzyme is as well characterized structurally at present at lysozyme, yet for ribonuclease (24–26), chymotrypsin (27, 28), carboxypeptidase (29–34), papain (35, 36) and human carbonic anhydrase (37), a molecular explanation of enzymic action is only a matter of time, and many other enzymes are beginning to be studied by x-ray diffraction. One of the enzymes most intensively studied by chemical methods is ribonuclease, whose sequence is shown below.

Ribonuclease is a chain-cutting enzyme like lysozyme and is very similar in its general construction. It is also an ovoid molecule with its active site in a crevice running across the waist of the molecule. It has 124 amino acids rather than 129 and a molecular weight of 13,700. Instead of cutting a polysaccharide chain, it cuts the polyribonucleotide chain shown to the right. This ribonucleic acid (RNA) is an alternating copolymer of ribose sugar and phosphate, with one of four kinds of organic bases attached to each ribose. These can be purine rings such as adenine (A) and guanine (G), or pyrimidine rings such as cytosine (C) and uracil (U). RNA, of course, is the raw material of the genetic messenger coming from the DNA of the nucleus and also the material of the ribosomes at which transcription of the messenger sequence takes place. Ribonuclease is secreted in the pancreas and is used to digest RNA in the duodenum. It cuts RNA in a specific manner, by cleaving the P—O bond on the far side of a phosphorus which is connected to the 3′ carbon of a ribose ring that bears a pyrimidine (C or U). These cleavage points are indicated at the right.

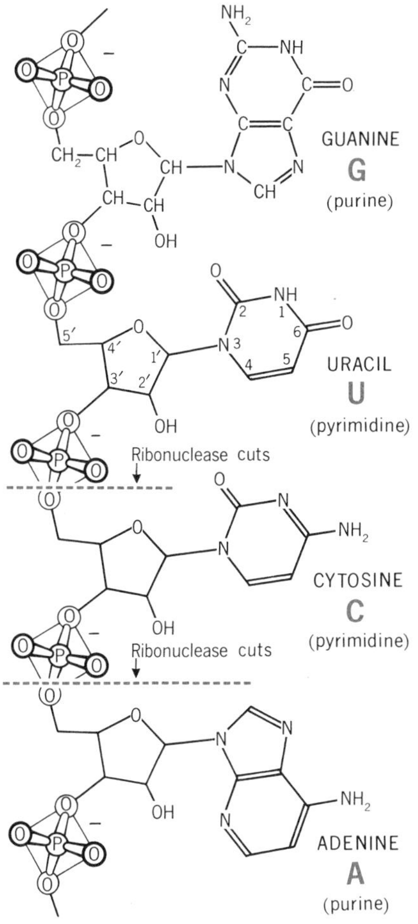

GUANINE G (purine)

URACIL U (pyrimidine)

Ribonuclease cuts

CYTOSINE C (pyrimidine)

Ribonuclease cuts

ADENINE A (purine)

Ribonucleic-acid chain (RNA).

1 lys glu thr ala 5 ala ala lys phe glu 10 arg gln his met asp 15 ser ser thr ser ala 20 ala ser ser ser asn 25 tyr cys (↓ 84) asn

gln 30 met met lys ser arg 35 asn leu thr lys asp 40 arg cys (↓ 95) lys pro val 45 asn thr phe val his 50 glu ser leu ala asp val

55 gln ala val cys (↓ 110) ser 60 gln lys asn val ala 65 cys (↓ 72) lys asn gly gln 70 thr asn cys (↓ 65) tyr gln 75 ser tyr ser thr met 80 ser ilu

thr asp cys (↓ 26) 85 arg glu thr gly ser 90 ser lys tyr pro asn 95 cys (↓ 40) ala tyr lys thr 100 thr gln ala asn lys 105 his ilu ilu val

110 ala cys (↓ 58) glu gly asn pro 115 tyr val pro val his 120 phe asp ala ser 124 val

AMINO-ACID SEQUENCE OF BOVINE RIBONUCLEASE. *Hydrophobic residues—Val, Leu, Ilu, Met, Phe—are tinted grey. Hydrogen-bonding side chains—Ser, Thr, Gln, Asn—are tinted in color. Cys disulfide-bridge participants are outlined in black and their connections are indicated. Ribonuclease is the first protein to be completely synthesized from its amino acids (26a). The final product was enzymatically indistinguishable from native ribonuclease.*

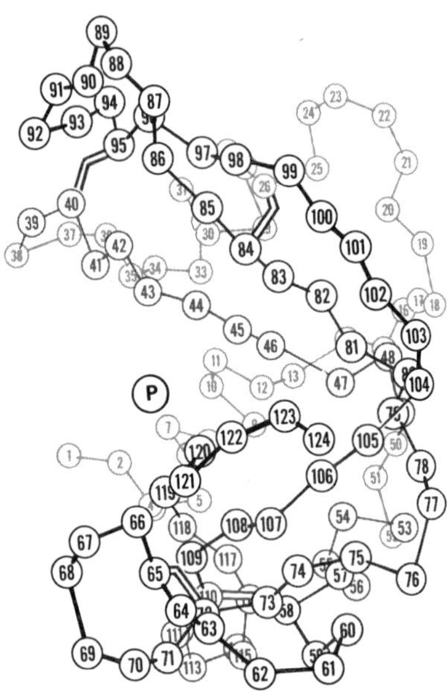

RIBONUCLEASE A. *The phosphate group in crevice opening (P) marks the active site. Disulfide bridges are represented only by a bent double line. Note the exposed chain to the upper right, which is cut to form ribonuclease S. (Coordinates by courtesy of Drs. G. Kartha and D. Harker, Roswell Park.)*

Ribonuclease and lysozyme perform quite different chemical acts, yet from the viewpoint of design of an enzyme they are similar; both cleave long chains in the middle. By analogy with *peptidase* and *nuclease*, a member of this general class of enzymes can be called a *catenase*, or chain-cutting enzyme. It cuts the chain in the middle rather than chewing away at one end, so it is more precisely an *endocatenase*.

The structure of the enzyme (left) has been worked out by Harker and Kartha (24). The structure of a more easily crystallizable variant in which the polypeptide chain has been cleaved between residues 20 and 21, ribonuclease S, has been solved by Richards and Wyckoff (25, 26). As the ribonuclease S analysis is presently at a more advanced stage, this structure (below) will be discussed in more detail. When the two forms are compared, it is apparent that, aside from the fanning out of the cut ends of the chain in ribonuclease S, there is little difference between them. The S peptide (residues 1–20) does not fall away from the rest of the molecule in ribonuclease S. It is primarily an α helix with a hydrophobic side and remains bound to the S protein by hydrophobic interactions, even though there is no longer a covalent connection. The chain cleavage is so harmless that ribonuclease S is still enzymatically active.

Ribonuclease and lysozyme are strikingly similar in their construction. The first regular features starting from the N-terminal end of the ribonuclease chain are two short lengths of α helix, 2–12 and 26–33, packed reasonably close to one another. As expected, these helices have hydrophobic sides facing the interior of the molecule. At residue 42, ribonuclease embarks upon a β-sheet structure but on a more extensive scale than in lysozyme. The first strand

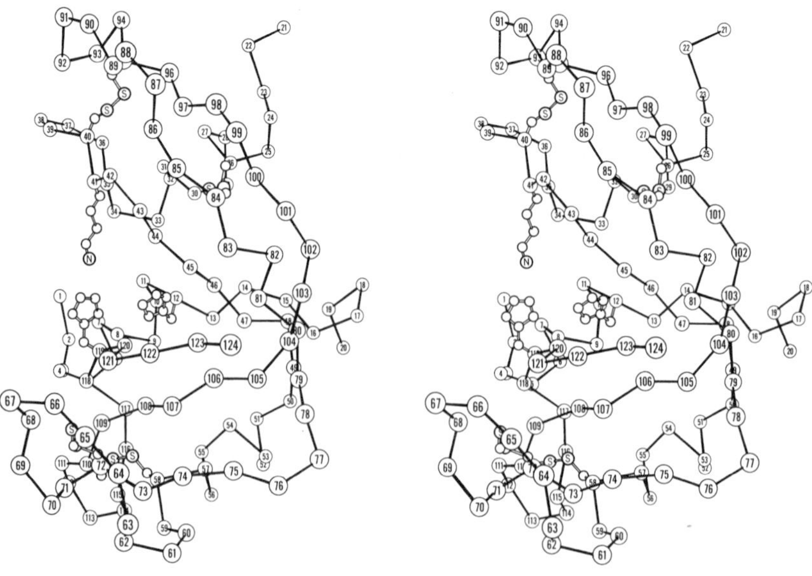

RIBONUCLEASE S IN STEREO

The two cut ends of the severed chain (20 and 21) have swung apart at the right, but there are few other changes from ribonuclease A. (Coordinates by courtesy of Drs. F. M. Richards and H. G. Wyckoff, Yale.)

of the sheet is laid down by 42–49, before the third and last length of α helix, 50–58. The chain then doubles back to begin the most striking feature of the molecule, a double-stranded V of β sheet which runs the entire length of the molecule, first up with residues 71–92 and then back down with 94–110, with 80–86 lying alongside the earlier strand 42–49. This V framework essentially defines the shape of the molecule and its crevice active site. The last act of the folding process is to bring the tail, 116–124, across the inside of the crevice and to position the catalytically important His 119 at the active site.

Like lysozyme, ribonuclease has a hydrophobic core on one side of the active site crevice and a less substantial second wing. Correlating well with the relative amounts of β structure, a full 34 percent of its residues are the potentially hydrogen-bridging Ser, Thr, Asn, or Gln, as compared with 25 percent for lysozyme and only 12 percent in myoglobin with no β sheet. These four residues are liberally distributed along the extended β chains except where a chain is surrounded by other chains and becomes hydrophobic, as 106–108.

The active site has been located, as in lysozyme, by binding an inhibitor to the enzyme and locating the inhibitor molecule in the map of the crystal. The inhibitor used was an analogue of a monomer of RNA, uridine 2′,3′-phosphate. Chemical studies have implicated His 12, His 119, and Lys 41 in the catalytic action, and all three show up close to one another at the active site. The mechanism of catalysis is not yet known, but x-ray analysis is providing a framework for discussion and a stimulus to new chemical work, just as with lysozyme (38).

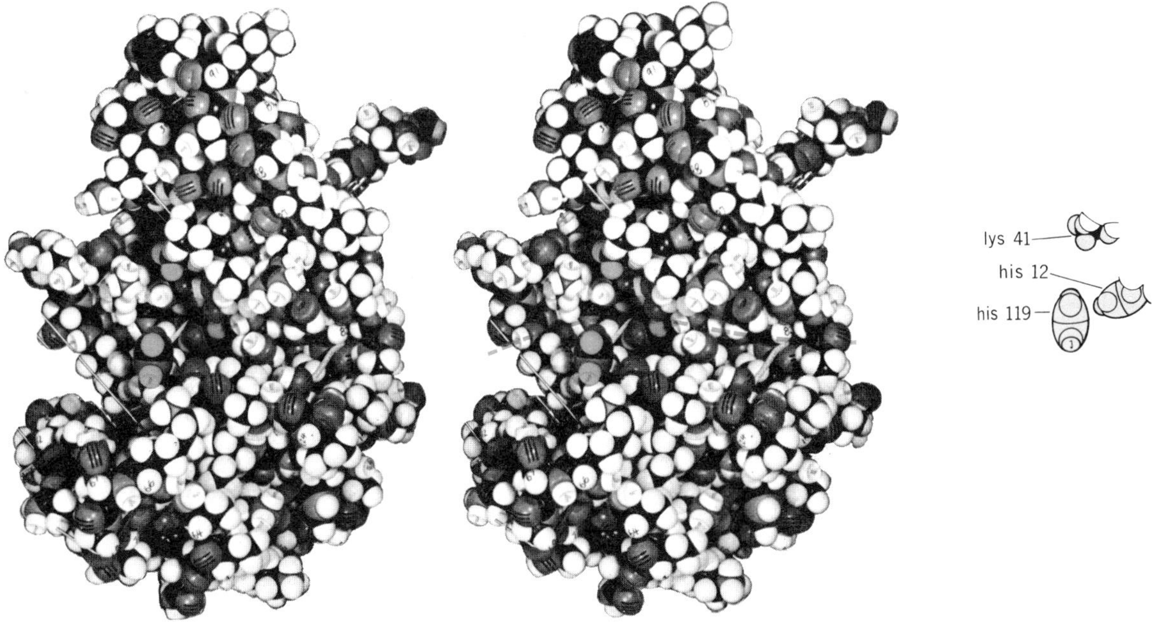

A more realistic space-filling model of ribonuclease S in stereo. The viewer is standing 45° to the left of the view in the stereo across the page, and is looking into the active site crevice, identified here by a dashed colored line. His 12, His 119, and Lys 41 are keyed below to clarify their position in the stereo pair, where they are shown in color.

4.4 THE PROTEASES: CHYMOTRYPSIN AND PAPAIN

A very interesting class of catenases is that of the enzymes which cut polypeptide chains, because enzymes themselves are polypeptide chains. The digestive enzymes of the stomach—pepsin and rennin—and pancreas—trypsin, chymotrypsin, elastase, and the carboxypeptidases—are best known chemically, but other proteases are found elsewhere, such as papain in the fruit of the papaya. These enzymes naturally cannot be synthesized at the ribosomes as such, for they are self-destructive. Instead, they are synthesized as inactive precursors or zymogens, which are converted to the active enzyme just before use by acid or by another enzyme.

Pepsinogen (mol. wt. 42,000) is cleaved by acid or by pepsin to form a pepsin/inhibitor complex (38,000) and five small peptides (4000 total). The free pepsin molecule has a molecular weight of 35,000. Trypsinogen, of molecular weight near 24,000, loses its first six residues to become active trypsin. The conversion is triggered in the small intestine by a special enzyme, enterokinase, but then goes extremely rapidly as the product trypsin acts as an activator. Chymotrypsinogen A becomes an active enzyme by a complicated series of steps (28) which can best be followed with the aid of the zymogen sequence as given on page 84. The zymogen is converted to π-chymotrypsin, an active enzyme, merely by the splitting by trypsin of the Arg 15—Ilu 16 peptide bond. Further treatment by chymotrypsin itself cleaves the Leu 13—Ser 14 bond and removes the dipeptide Ser 14—Arg 15 to form δ-chymotrypsin, and still further chymotrypsin digestion removes another dipeptide, Thr 147—Asn 148, to form the working molecules α- and γ-chymotrypsin. These latter are only the low- and high-pH forms of the same protein (39). The inactive precursor procarboxypeptidase A is a complex assembly of three subunits with a molecular weight of around 88,000. It is split by trypsin in an involved set of reactions (40), and only one of the three subunits eventually becomes the 34,600 molecular weight carboxypeptidase A.

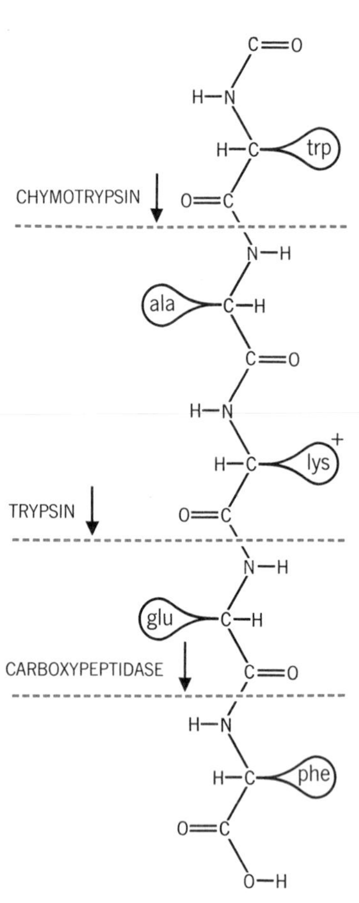

A polypeptide chain, with the cleavage points for chymotrypsin, trypsin, and carboxypeptidase marked.

The characteristic chain-cleavage points for these enzymes are shown to the left. Chymotrypsin and pepsin cut in the middle of a polypeptide chain, severing the peptide bond just beyond the carbonyl group of a residue with an aromatic side chain; trypsin acts similarly with positively charged side chains. Papain is also an endopeptidase but is much less selective. The carboxypeptidases are exopeptidases, carboxypeptidase A removing a C-terminal residue most easily when it is aromatic, and carboxypeptidase B functioning best when the last residue is basic, reminiscent of the chymotrypsin/trypsin complementarity.

Because these digestive enzymes had been so well purified and crystallized, and so extensively studied by chemists, they were natural candidates for x-ray work. All the enzymes mentioned above are in one stage or another of crystal structure analysis,* and structural results have been obtained for chymotrypsin, elastase, subtilisin BPN′, papain, and carboxypeptidase.

Chymotrypsin, trypsin, elastase, thrombin, subtilisin, phosphorylase, alkaline phosphatase, and many other enzymes are serine proteases or esterases. These enzymes have a particularly reactive serine at their active site which is necessary for catalytic activity. The mechanisms that have been proposed for these enzymes usually have Ser attacking a bond by donating a proton, much as Glu 35 did in Phillips' mechanism for lysozyme. The essential serine in chymotrypsin is Ser 195 and is the equivalent Ser 183 in trypsin. (The sequence comparison on page 84 uses only the chymotrypsin numbering.) His 57 (in the chymotrypsin numbering) is also necessary in both proteins. The two enzymes are quite similar in their origin, rate of synthesis, endocatenase activity, mode of catalytic action (41, 42), and, as the sequence comparison shows, in amino acid sequence. They differ chiefly in substrate specificity: Trypsin cuts the polypeptide chain on the carbonyl side of a basic residue, Lys or Arg, whereas chymotrypsin cuts most easily those polypeptide bonds on the carbonyl side of a large aromatic residue (left). The sequence comparison suggests some sort of family relationship between the two enzymes, and the x-ray analysis has provided striking confirmation of this.

A preliminary structure analysis of α-chymotrypsin has produced the polypeptide chain structure on page 85 (27, 28). The molecule is roughly spherical, with many parallel extended chains that are probably of β structure and with only two short

* Pepsin by N. Andreeva of the Academy of Science in Moscow; rennin by C. Bunn at the Royal Institution in London; elastase by B. Hartley, D. M. Shotton, and H. C. Watson in the Medical Research Laboratories in Cambridge and Bristol Universities; and trypsin by R. Stroud, L. Kay, and R. E. Dickerson at Caltech. Others as noted.

PROPOSED CHYMOTRYPSIN MECHANISM

The interaction with His 57 and with a buried Asp 102 (right) gives Ser 195 a negative charge and makes it a strong nucleophile, suggesting the cleavage mechanism shown in the four steps to the right above (28a). Colored arrows mark shifts of electrons. An identical hydrogen-bonding network between Ser, His, and a buried Asp has been found in elastase, an evolutionary cousin with identical chain folding (28b). But it has also been found in subtilisin BPN′ (28c), another protease with similar catalytic activity but otherwise quite different structure. For these side chains in their molecular setting, and for a similar view of the active site of subtilisin, see the stereo supplement.

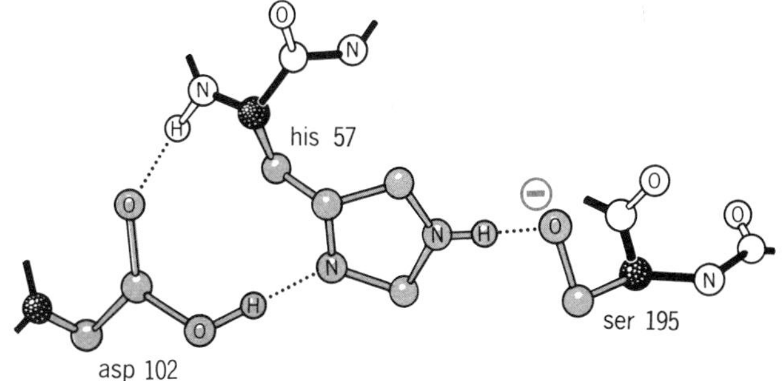

	1 (122)				5					10			13	14	15	16				20					25			
C	cys	gly	val	pro	ala	ilu	gln	pro	val	leu	ser	gly	leu	ser	arg	ilu	val	asn	gly	glu	glu	ala	val	pro	gly	ser	trp	pro
T	.....	.....	.....	.....	.....	.....	.....	.....	.....	val	asp	asp	asp	asp	lys	ilu	val	gly	gly	tyr	thr	cys	gly	ala	asn	thr	val	pro
																						157						

		30					35					40					45					50					55	
C	trp	gln	val	ser	leu	gln	asp	lys	thr	gly	phe	his	phe	cys	gly	gly	ser	leu	ilu	asn	glu	asn	trp	val	val	thr	ala	ala
T	tyr	gln	val	ser	leu	asn	.....	.....	ser	gly	tyr	his	phe	cys	gly	gly	ser	leu	ilu	asn	ser	gln	trp	val	val	ser	ala	ala
														58														

				60					65					70					75					80				
C	his	cys	gly	val	thr	thr	ser	asp	val	val	val	ala	gly	glu	phe	asp	gln	gly	ser	ser	ser	glu	lys	ilu	.....	gln	lys	leu
T	his	cys	tyr	lys	ser	.....	gly	ilu	gln	val	arg	leu	gly	gln	.....	asp	asn	ilu	asn	val	val	glu	gly	asn	gln	gln	phe	ilu
		42																										

		85					90					95					100					105					110	
C	lys	ilu	ala	lys	val	phe	lys	asn	ser	lys	tyr	asn	ser	leu	thr	ilu	asn	asn	asp	ilu	thr	leu	leu	lys	leu	ser	thr	ala
T	ser	ala	ser	lys	ser	ilu	val	his	pro	ser	tyr	asn	ser	asn	thr	leu	asn	asn	asp	ilu	met	leu	ilu	lys	leu	lys	ser	ala

			115						120		1			125					130						135			
C	ala	ser	phe	ser	gln	thr	val	ser	ala	val	cys	leu	pro	ser	ala	ser	asp	asp	phe	ala	ala	gly	thr	thr	cys	val	thr	thr
T	ala	ser	leu	asn	ser	arg	val	ala	ser	ilu	ser	leu	pro	thr	ser	cys	ala	.....	.....	ser	ala	gly	thr	gln	cys	leu	ilu	ser
																232									201			

	140					145	146	147	148	149	150					155					160					165		
C	gly	trp	gly	leu	thr	arg	tyr	thr	asn	ala	asn	thr	pro	asp	arg	leu	gln	gln	ala	ser	leu	pro	leu	leu	ser	asn	thr	asn
T	gly	trp	gly	asn	thr	lys	ser	ser	gly	thr	ser	tyr	pro	asp	val	leu	lys	cys	leu	lys	ala	pro	ilu	leu	ser	asn	ser	ser
																		22										

			170					175					180					185							190			
C	cys	lys	lys	tyr	trp	gly	thr	lys	ilu	lys	asp	ala	met	ilu	cys	ala	gly	ala	.....	.....	ser	gly	val	ser	ser	cys	met	gly
T	cys	lys	ser	ala	tyr	pro	gly	gln	ilu	thr	ser	asn	met	phe	cys	ala	gly	tyr	leu	glu	gly	gly	lys	asn	ser	cys	gln	gly
	182														168											220		

		195					200					205					210					215					220	
C	asp	ser	gly	gly	pro	leu	val	cys	lys	lys	asn	gly	ala	trp	thr	leu	val	gly	ilu	val	ser	trp	gly	ser	ser	thr	cys	ser
T	asp	ser	gly	gly	pro	val	val	cys	ser	gly	.....	.....	.....	.....	lys	leu	gln	gly	ilu	val	ser	trp	gly	ser	gly	.....	cys	ala
								136																			191	

					225					230					235					240					245
C	thr	ser	.....	thr	pro	gly	val	tyr	ala	arg	val	thr	ala	leu	val	asn	trp	val	gln	gln	thr	leu	ala	ala	asn
T	gln	lys	asn	lys	pro	gly	val	tyr	thr	lys	val	cys	asn	tyr	val	ser	trp	ilu	lys	gln	thr	ilu	ala	ser	asn
												127													

BOVINE CHYMOTRYPSINOGEN

Amino-acid sequences of bovine chymotrypsinogen (C) and trypsinogen (T), arranged for maximum homology.

Hydrophobic residues (including Ala when homologous with another hydrophobic residue).

Hydrogen-bonding side chains

Disulfide-bridge participant

..... *Gap in sequence*

Cleavage points in activation of zymogen.

segments of α helix at residues 164–170 and at 234–245. The catalytic site is a shallow depression with residues 57, 120, and 195, next to a hydrophobic pocket (right) which gives the enzyme its specificity. Preliminary evidence (43) suggests that chymotrypsinogen and all the various chymotrypsins have basically the same folding, and that activation involves removal of the two dipeptides without any gross changes in the protein molecule. The cut ends of the three chains in α-chymotrypsin are visible in the drawing as 13 . . . 16 and 146 . . . 149. Blow and coworkers (28) have proposed that, upon activation, the newly formed amino group of Ilu 16 attracts the carboxyl side chain of Asp 194 and causes it to swing out of the way and unblock the active site. In a similar way, a less active form is proposed to occur from the α form at high pH when the Ilu 16 amino group is deprotonated and the carboxyl of Asp 194 swings back across the active site.

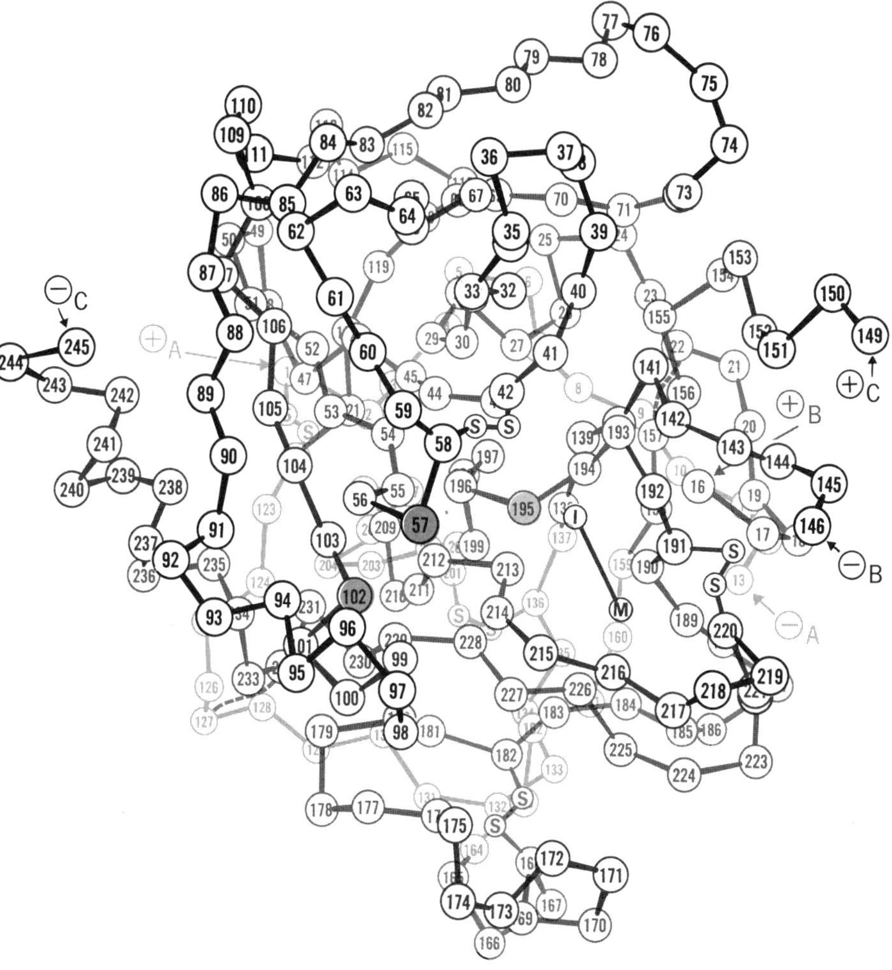

α CHYMOTRYPSIN

Folding of the main chain with chymotrypsinogen numbering. ⊕ *and* ⊖ *mark the amino-and carboxyl-terminal ends of the A, B, and C chains.*

M and I locate the methyl and sulfur groups of the tosyl inhibitor: CH_3—C_6H_4—SO_2— *at the active site. The methyl end fits into the hydrophobic pocket formed by 184–191 and 214–227.*

Note the suggestion of a twisted sheet in the strands of residues 91–86, 103–108, 55–50, 39–46, and 35–29. The "vestigial" disulfide connections, present in trypsin but not in chymotrypsin, are between residues 22–157 and 127–232. The extra dipeptide in trypsin, which contains the Glu that may be responsible for the preference of trypsin for the basic side chains, is inserted between residues 185 and 186. The catalytically important His 57, Asp 102, and Ser 195 are in color. (Coordinates by courtesy of Dr. D. M. Blow, Cambridge, England.)

More will undoubtedly be learned about chymotrypsin action as x-ray analysis and chemistry progress in tandem. But one other striking aspect of the structure must be mentioned. Neurath and coworkers first pointed out the extensive parallelism between the amino acid sequences of chymotrypsin and trypsin. Four of the five disulfide bridges of chymotrypsin occur at the same places along the chain as four of the six trypsin bridges. All the essential residues match, as do a great many others. Forty-one percent of the chain is identical in the two proteins. The unique chymotrypsin disulfide bridge, 1—122, cannot exist in trypsin because trypsin has lost the initial tail. (Note that the activation cleavage points match, so the beginning of the trypsin chain at Ilu 16 is the same as the beginning of the *second* chymotrypsin chain after activation.) But the two bridges peculiar to trypsin tell an interesting story. In the chymotrypsin sequence, they are 22–157 and 127–232.

They must ultimately be found to be close in trypsin, when its structure is worked out. But although there is no ad hoc reason to expect these now *unlinked* pairs of residues to be neighbors in chymotrypsin, x-ray analysis has shown that they are. Two explanations are possible, both with the same implications for the evolutionary relatedness of these enzymes. One is that the original ancestor of both chymotrypsin and trypsin had such disulfide bridges. After the enzyme gene doubled, the two enzymes each went their own way, evolving the ability to cut up protein in different places. These two disulfide bridges were not essential to the folding of the molecule, so chance mutations at these loci in the DNA led to the elimination of the Cys in chymotrypsin but not yet in trypsin. The near approaches of residues 22 and 157, and 127 and 232 in chymotrypsin are now vestigial features, like the vermiform appendix in man or the pineal gland "third eye" of vertebrates. The alternative explanation proposes that chance mutations in the trypsin gene eventually led to two Cys residues facing one another across a close interchain approach. The disulfide bond formed by reduction then gave the molecule a sufficient advantage in stability to fix the mutations permanently in the population. Either explanation leads to the same conclusion: Chymotrypsin and trypsin are cousins.

Even more striking than the similarities of chymotrypsin, trypsin, and elastase is the discovery that the *same* active site structure (page 83) has evolved independently in subtilisin BPN′ (28c), a protease found in soil bacilli, although the folding of the rest of the molecule is very different. This is the clearest example yet of convergence of function, or the independent evolution of the same mechanism in unrelated enzyme molecules.

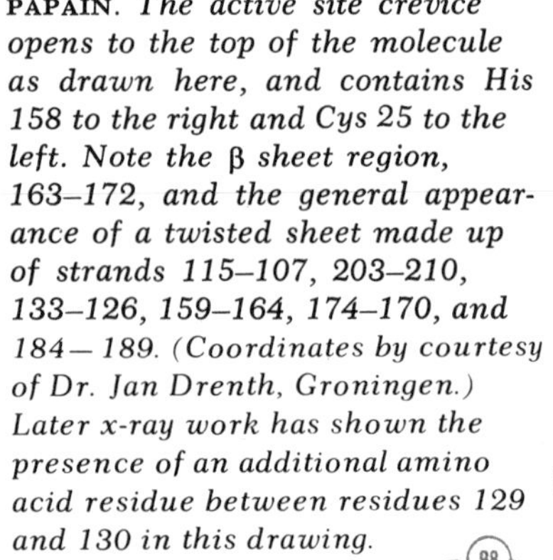

PAPAIN. *The active site crevice opens to the top of the molecule as drawn here, and contains His 158 to the right and Cys 25 to the left. Note the β sheet region, 163–172, and the general appearance of a twisted sheet made up of strands 115–107, 203–210, 133–126, 159–164, 174–170, and 184–189. (Coordinates by courtesy of Dr. Jan Drenth, Groningen.) Later x-ray work has shown the presence of an additional amino acid residue between residues 129 and 130 in this drawing.*

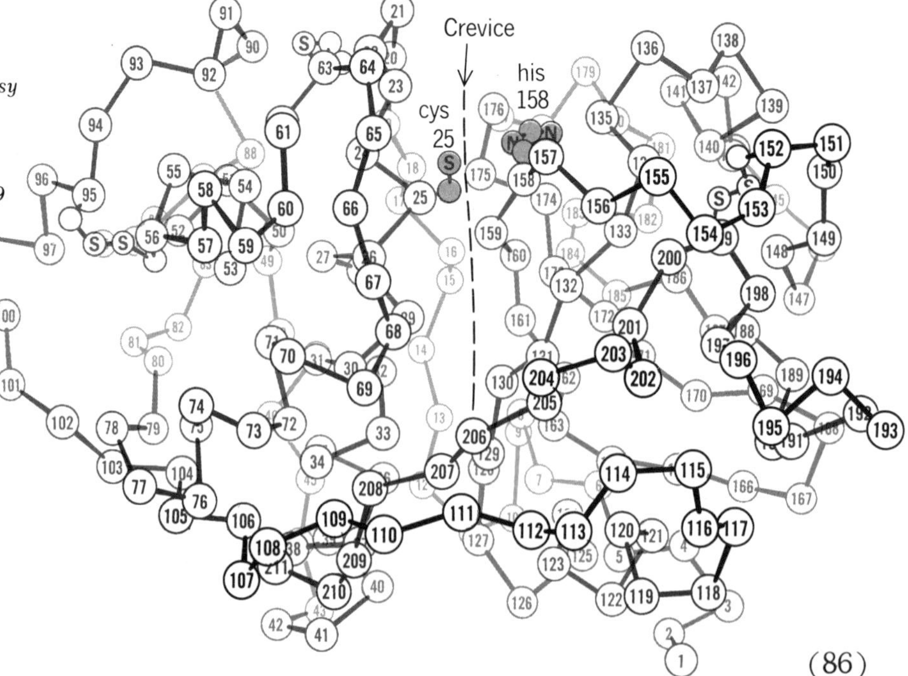

Papain is a sulfhydryl rather than a serine protease, with molecular weight 23,000, 211 amino acid residues in one continuous chain, and three disulfide bridges. Its structure is shown at the left below. It resembles lysozyme and ribonuclease in shape, with the active site being in a deep crevice across the molecule. The essential Cys 25 and His 158 face one another across the crevice. There are only four short stretches of α helix: 26–41, 50–56, and 69–78 in one wing of the molecule and 116–126 in the other. This latter wing also contains a lone portion of β structure, 163–172, and a hydrophobic core in which 12–14 side chains of Leu, Ilu, Val, and Phe are packed together.

Papain illustrates the advantage of cooperation between chemist and crystallographer in sequencing a protein. It is sometimes easier to work out the sequences within peptide fragments in a sequence analysis than it is to order the peptides properly to form the full protein chain, especially if it is not possible using different cleaving enzymes to obtain overlaps of peptides along the entire chain length. The sequence arrived at by chemical means (44) may be designated as A-B-C-D, representing four successive chain runs of 28, 107, 41, and 22 amino acids, respectively. The missing residues were acknowledged but unexplained. X-ray analysis (36) showed that the true chain order was A-X-C-B-D, with X being a new 13-residue stretch of chain. X-ray analysis can provide the overall picture with sometimes blurred details; chemical sequence analysis can provide the details but can go wrong in the broad picture. Clearly, an insistence upon either method alone for sequence analysis is unwise.

4.5 AN EXOCATENASE: CARBOXYPEPTIDASE A

The catenases that have been examined so far seem to form a class of similar enzymes: ovoid molecules with active site in a deep crevice, a two-wing structure with one wing forming a hydrophobic core and the other wing more open, limited use of relatively short lengths of α helix, greater or lesser amounts of exposed β structure, and "random chain" (a misleading expression meaning only that the chain folding is too complex for us to understand and pigeonhole. After as much as 3000 million years of natural selection, nothing in a protein is random). Chymotrypsin fits the pattern least well, but the extensive sequence homologies among chymotrypsin, trypsin, and elastase (40) suggest that a second pattern is emerging. But the last enzyme of this chapter, carboxypeptidase A, is strikingly different from anything we have seen so far. The β sheet is inside, and the α helices are out. There is little "random chain" except at one point, and the β sheet is really a sheet, being eight chains wide.

Carboxypeptidase A is just what its name implies: an enzyme that digests polypeptide chain from the carboxyl terminal end. It cleaves away the last residue most easily when this residue has a large aromatic side group, and in this respect its specificity resembles that of chymotrypsin. It and the several aminopeptidases can be classed as exocatenases.*

The high-resolution map of carboxypeptidase is very clear (29–34), and Lipscomb's group has come close to being able to sequence the protein from x-ray analysis. The chemical sequence work (45) has proved to be especially difficult, but a combination of the two methods should lead to a correct and unambiguous sequence. The molecule is roughly spherical, 52 by 44 by 40 Å, with 307 amino acid residues and a molecular weight of 34,600. It has one continuous polypeptide chain, with one disulfide bridge and with a Zn atom that is necessary for catalytic activity (although it can be removed and replaced with several transition metals). The Zn atom sits in a broad depression in the surface of the molecule, which in fact is the active site, and from which a pocket extends into the interior of the molecule. The structural heart of the molecule is a β sheet made up of no less than eight parallel or antiparallel extended chains, as shown to the right. Roughly 20 percent of the residues are involved in this massive pleated sheet, which is twisted about a vertical axis so that the top chain (η) is rotated by 120° with respect to the bottom chain (α) (below). On the two sides of this twisted sheet are packed eight α helices of various lengths and degrees of perfection, which employ another 35 percent of the residues. A single "random-chain" region (in the sense of page 87) uses another 20 percent, and the remaining residues occur in the chain segments needed to tie these features together.

* Lipscomb has called attention to this additional example of the illogical nature of the English language. Carboxypeptidase cuts residues from the *end* of the chain; therefore, it is *not* an endopeptidase!

The main structural elements of carboxypeptidase A, seen as in the large drawing on page 90. Left: eight-chain twisted β sheet. Center: β sheet packed with α helices. Right: addition of the extended chain that molds the shape of the active site depression.

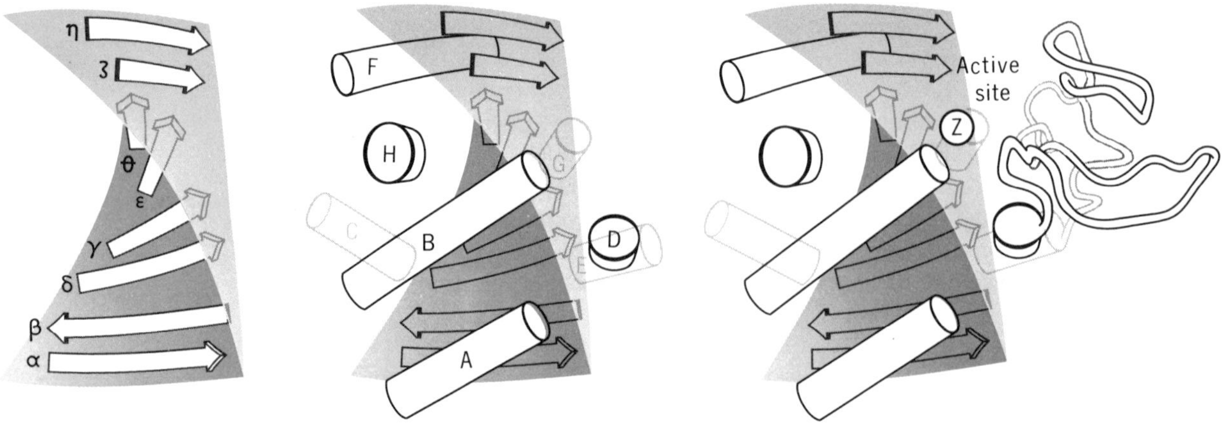

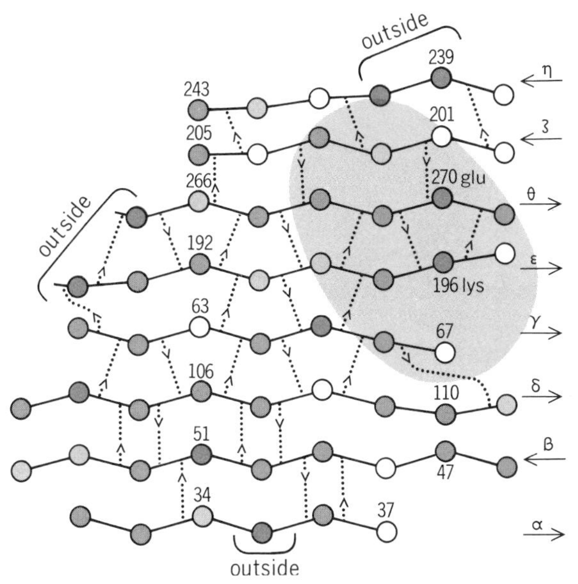

Active site region (on far side of sheet)

Hydrophobic: Ala, Val, Leu, Ilu, Met, Phe, Pro

Ser or Thr

Charged polar: Asp, Glu, Lys, Arg

Miscellaneous

Hydrogen bond

(C=O······H—N)

The β sheet core of carboxypeptidase A. Residues pointing up on a zigzag chain extend out to the side of the sheet that contains the active site (into the paper in this view). Residues bent down extend out of the paper as drawn here, or to the "left" side of the sheet as drawn on page 90. Note that all charged groups in the sheet are on the outside except those involved in the active site, and that the sheet itself is generally hydrophobic.

The sequential order of these structural components is shown in the table to the lower right. The sections of α helix have been lettered A–H for the purposes of this chapter, and the individual strands of the β sheet have been given the symbols α–θ, again in sequential order in the polypeptide chain. Note that this is *not* the order of the chains across the sheet. Starting from the N-terminal end of the polypeptide chain, after laying down helix A and strands α and β of the sheet, the chain then skips one row in the sheet and makes γ the *fourth* strand, forms helices B and C, and only then lays down strand δ *between* β and γ. The chain then generates helix D, the "random" region, helix E, and sheet strand ε. But then, unexpectedly, it skips a strand once more and lays down ζ two chains up from ε, forms helix F, the topmost sheet strand η, and helix G. It drops down from the top of the molecule to fill in the missing strand θ between ε and ζ, and concludes the molecule with the long helix H. The enzyme protects its own C-terminal end against self-digestion by tucking the final Asn 307 back inside with a salt link to Arg 265. It is difficult to believe that this elaborate procedure represents the way the chain really folds at the ribosome. It is more likely that the helical regions coil first and that their packing in some way generates the β sheet, which then gives the finished molecule added stability. From experience with other proteins, we would expect an interior β sheet to be made up primarily of *hydrophobic* residues and the α helices to pack hydrophobic sides to the sheet. This, it turns out, is true. The molecule is generated not as the enzyme but as the much larger procarboxypeptidase, so the other two units of the proenzyme may play some role in ensuring that the part that becomes carboxypeptidase is folded properly.

α Helix	β Sheet	Residues
A		14 . . . 29
	α	32 . . . 37
	β	46 . . . 54
	γ	61 . . . 67
B		72 . . . 88
C		94 . . . 103
	δ	103 . . . 111
D		115 . . . 122
"Random chain"		122 . . . 174
E		174 . . . 184
	ε	190 . . . 197
	ζ	200 . . . 205
F		215 . . . 233
	η	238 . . . 243
G		254 . . . 262
	θ	265 . . . 271
H		288 . . . 305

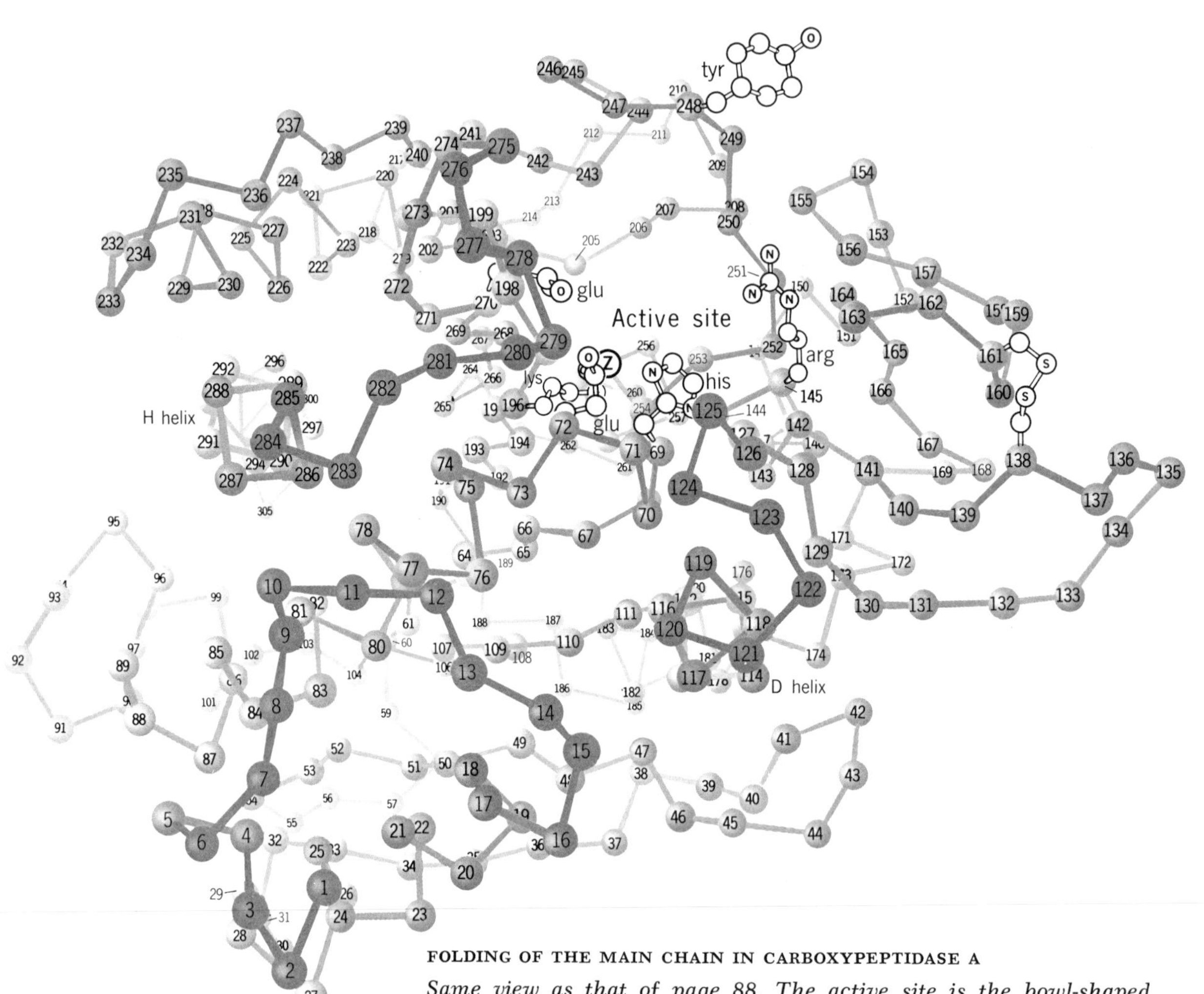

FOLDING OF THE MAIN CHAIN IN CARBOXYPEPTIDASE A

Same view as that of page 88. The active site is the bowl-shaped depression at the upper right, below Tyr 248. The groove down the face of the molecule, into which the remainder of the polypeptide chain is presumed to fit, runs between the shoulders formed by 277–281 and 123–127. The zinc atom (Z) is shown. The two **Cys**, *138 and 161, are joined in a disulfide bridge. The active side chains Tyr 248, Glu 270, and Arg 145 are shown in their positions in the absence of substrate or inhibitor. Helices H (left) and D (right) are seen nearly end-on. Note how the extended chain, 124–174 and 241–251, loops back again and again to shape the active site. Both the geography of the active site and the catalytic mechanism (page 92) resemble those of chymotrypsin. Both enzymes have an active-site depression, with substrate specificity produced by a nearby hydrophobic pocket into which an aromatic side chain of the polypeptide chain being cut is inserted. In both enzymes, the carbonyl carbon of the peptide bond being cut forms a temporary covalent intermediate with a nucleophile: Ser 195 in chymotrypsin and Glu 270 in carboxypeptidase. And finally, in both enzymes, a neighboring group (His 57 or Tyr 248) donates a proton to the nitrogen of the bond being broken, and is replenished by a proton from water. (Coordinates by courtesy of Dr. W. N. Lipscomb, Harvard.)*

The arrangement of these structural components is shown in a "side" view of the molecule on the opposite page and in a "top" view in the stereo pair below. In the top view the β sheet spirals in a clockwise direction as it rises out of the page. Helices A, B, C, D, F, and H are packed on the left side of the sheet and give the molecule its basic solidity. Helices E and G and the "random" region build the right side of the molecule, which contains the active site. In the side view the chain sequence suggests that the molecule is built from the bottom to the top and then partway down again. It is difficult to imagine how helix H could be spun between B, C, and F, except by the packing together of preformed helices.

The β sheet is constructed differently than in lysozyme, depending only upon amide-carbonyl bonds for its integrity rather than on side chains of Ser, Thr, Asn, and Gln. In fact, most of the side chains are hydrophobic (page 89), as would be expected in a β sheet designed for the interior of a molecule. The α helices have hydrophobic sides which are packed against the sheet. There are enough regular cross-chain hydrogen bonds that one can describe the structure as a true β pleated sheet rather than merely "β-like folding." Furthermore, both the regular hydrogen bonds of the antiparallel β pleated sheet (page 34) and the skewed bonds of the parallel-chain pleated sheet (46) are seen. This marks the first time that the Pauling–Corey parallel-chain β pleated sheet has been unambiguously demonstrated in nature.

The zinc atom is tetrahedrally coordinated, with three of its four ligands coming from the nitrogens or oxygens of His 69, Glu 72, and Lys 196.* The fourth ligand is supplied by the substrate, in the form of the carbonyl oxygen of the peptide bond that is being split. Three other side chains around the edge of the active site appear to interact directly with the substrate and to swing to a new conformation when they do so: Arg 145, Tyr 248, and Glu 270. As the drawing to the left and the stereo below demonstrate, many of the elbows or corners of chain between one structural feature and the next are swung around to form the rim of the active site depres-

* Later chemical work has shown that residue 196 is His and not Lys.

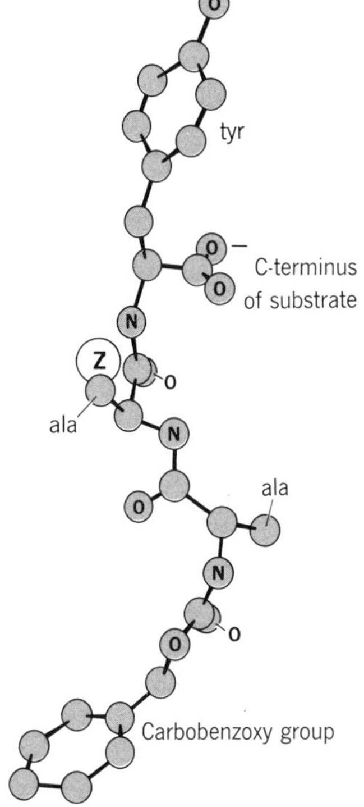

The carbobenzoxy-Ala-Ala-Tyr substrate of carboxypeptidase A. The substrate has been fitted to the enzyme by analogy with the observed Gly-Tyr inhibitor.

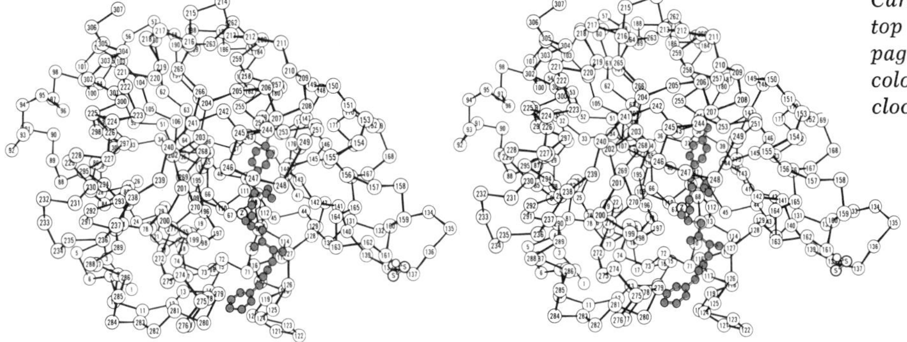

Carboxypeptidase A, seen from the top in the drawing on the opposite page. The substrate is shown in color. Note how the β sheet spirals clockwise as it rises out of the page.

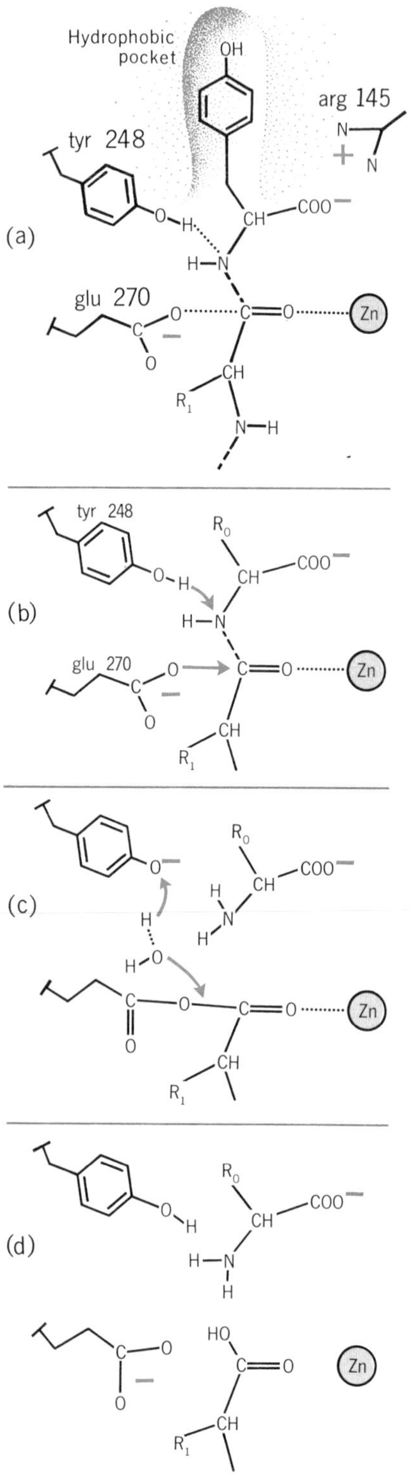

sion, and several of these contain the important side chains. The bend between sheet strand γ and helix B (residues 67–72) bears both His 69 and Glu 72. The longer loop between η and G (241–253) brings Tyr 248 to the active site. Lys 196 and Glu 270 are on the right-hand wall of the twisted sheet (page 89), which itself forms the floor of the active site depression. Arg 145 is positioned by the folding of the "random" region (123–174). The "random chain," in fact, defines the shape of the lower rim of the active site (right).

The substrate is shown in place in the drawing on the right, in a view looking directly into the active site depression. The positions of Arg 145, Tyr 248, and Glu 270 *before* the binding of substrate are given dashed outlines. The inhibitor molecule shown is

$$C_6H_5—CH_2—O—CO—[—NH—CH(CH_3)—CO—]_2$$
$$—NH—CH(CH_2—C_6H_4—OH)—COOH$$

or carbobenzoxy-Ala-Ala-Tyr. The x-ray work was actually done with the uninhibited enzyme and with a Gly-Tyr inhibitor. The path of the *left* half of the inhibitor on the opposite page was obtained from model building. A longer polypeptide chain is presumed to continue to the lower left along the groove flanked by the loops of chain carrying residues 279 and 125. The stereo on page 91 shows best how the aromatic C-terminal side group of the substrate fits into a pocket in the interior of the molecule, whose rim is the chain of residues 245–251. Arg 145 moves 2 Å closer to interact with the substrate's terminal carboxyl group. Tyr 248 swings 14 Å down to place its hydroxyl group near the nitrogen of the bond to be split, and the neighboring carbonyl oxygen coordinates with the Zn. Glu 270 moves near the carbonyl carbon.

One possible mechanism that has been suggested by x-ray analysis is given in the margin to the left. The inhibitor, and by inference the substrate, binds at the active site with its large aromatic side chain inserted into the deep hydrophobic pocket. The polypeptide chain may bind along the groove down the flank of the molecule and bring the next three side groups near a cluster of hydrophobic residues (Tyr 198, His 279). This could account for the preference of the enzyme not only for an aromatic final substrate residue but for hydrophobic residues in the final few positions also.

It is proposed (32) that the Glu 270 side chain acts as a nucleophile to form a transient anhydride with the carbonyl group of the substrate (a, b, left), while the Tyr proton attacks the nitrogen at the other end of the weakening peptide bond. The Zn atom both orients the substrate and polarizes the carbonyl group, aiding in the formation of the anhydride. A subsequent hydrolysis step restores the Tyr proton (c) and cleaves the anhydride to leave the peptide bond broken and the enzyme side groups in their original state (d).

The most probable mechanism of catalytic action.

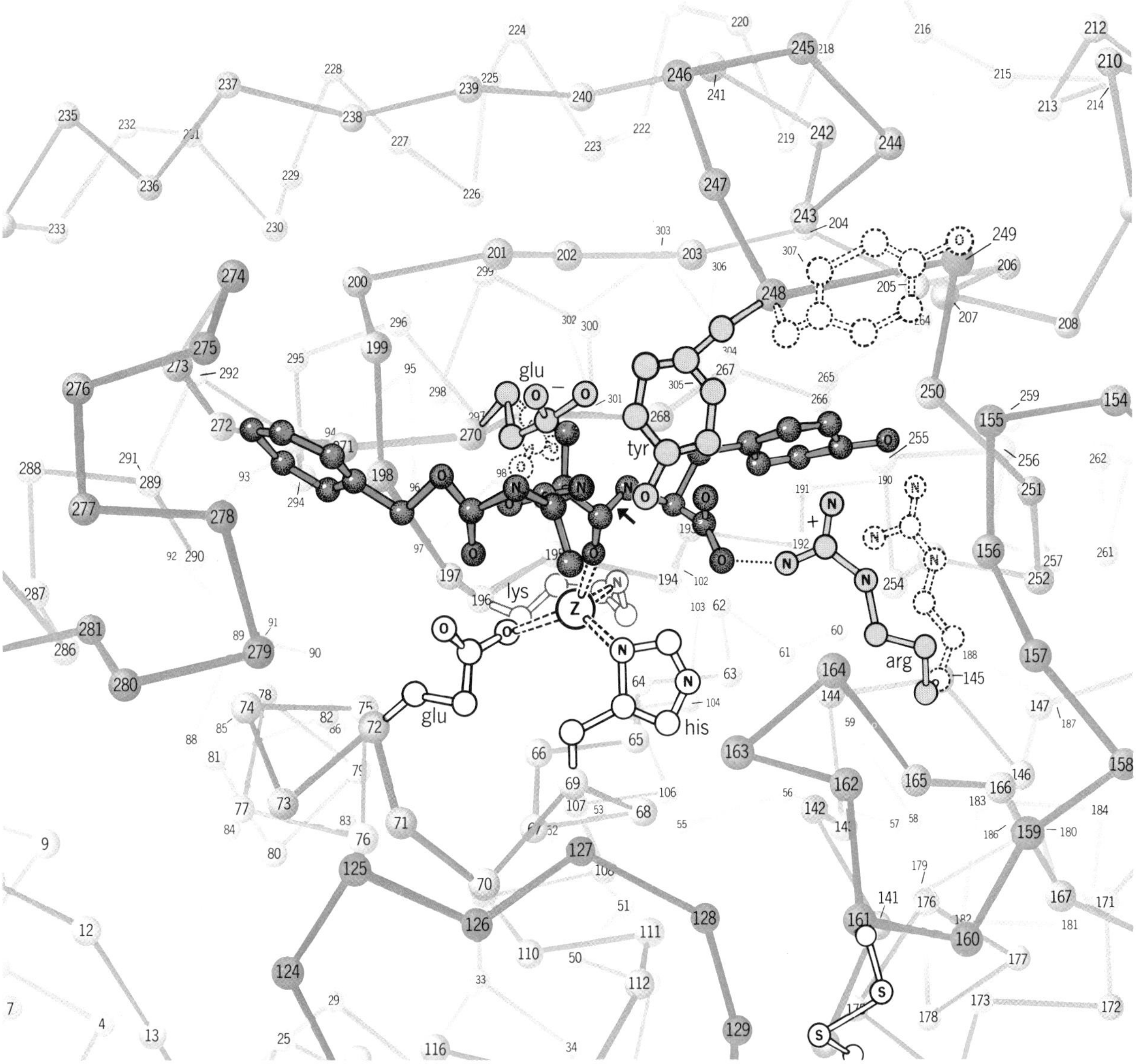

THE ACTIVE SITE WITH SUBSTRATE IN PLACE

Arg 145, Tyr 248, and Glu 270 are shown before substrate binding (dotted) and after (solid). Note the tetrahedral coordination of the zinc atom (Z) to His 69, Glu 72, Lys 196 (now known to be His 196), and the carbonyl oxygen of the bond to be cleaved in the substrate.

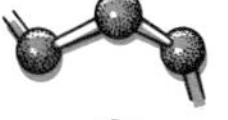 *Inhibitor*

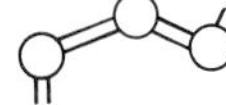 *Zn-liganding side chains*

 *Moving side chains* without *inhibitor*

 *Same,* with *bound inhibitor*

If this is not the precise mechanism, at least x-ray analysis has set boundary conditions on possible schemes, and it is the task of the enzyme chemist now to design and carry out the proper experiments that will lead to the exact mechanism.

The end result of any x-ray analysis of a protein is a map of electron density of the crystal in three dimensions. This is conveniently plotted in sections through the crystal, in contours of equal electron density, and transferred to plexiglass sheets that can be stacked to make the entire molecule visible at once. Two sets of carboxypeptidase sections are shown here.

This pattern of electron density in three dimensions must then be interpreted in terms of the arrangement of atoms in the molecule. The interpretation is greatly aided by the presence of a continuous polypeptide chain with prominent carbonyl oxygens and side groups recurring at regular intervals, and by the knowledge that each side group must be one of 20 known structures. The dots in these sections mark the positions of atoms deduced by best fitting groups of known stereochemistry to the electron density. Although it is possible in principle to sequence a protein from its x-ray structure, in practice the chemical sequence is a great aid in the rapid interpretation of the electron-density map.

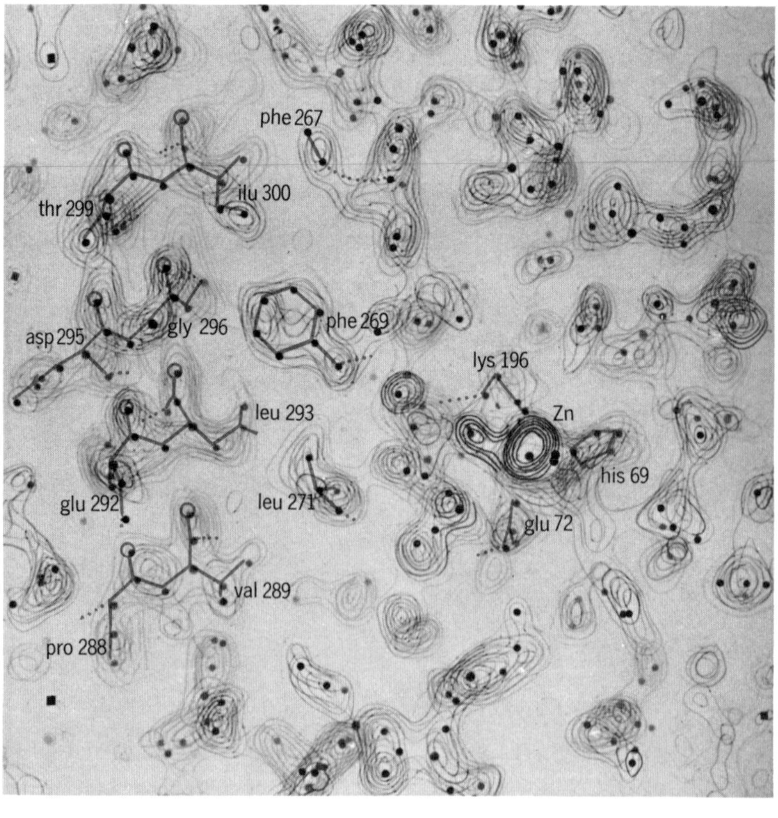

phe 267
thr 299
ilu 300
asp 295
gly 296
phe 269
lys 196
Zn
leu 293
his 69
glu 292
leu 271
glu 72
val 289
pro 288

Five sections through the 2.8 Å high-resolution electron density map of carboxypeptidase A, seen as in the stereo on page 91. The interpretation of polypeptide chain and side groups is overprinted in color. The zinc atom and its three ligands are at the lower right, and α helix H (residues 288–300) runs vertically along the left side. The easily identifiable main chain carbonyl groups are marked: —○ Phe 267, Phe 269, and Leu 271 extend down into these sections from β chain strand θ, which is visible in the sections on the opposite page.

4.6 ENZYME ARCHITECTURE

To date two of the three basic fibrous protein structures have been found in globular proteins: the right-handed α helix and both forms of the β pleated sheet. So far the collagen triple helix has not appeared, but after carboxypeptidase it is a rash man who would be dogmatic about the future. No left-handed helix of any kind has been encountered, which is not surprising in view of the clash of side groups and carbonyl oxygens in L-amino acids. Some 3_{10} helix has been found, occasionally alone but more often as the final turn of an α helix. The α helices tend to have one hydrophobic and one hydrophilic side, and the β-like structures, when exposed, are made up mostly of polar groups with an especially high concentration of Ser, Thr, Asn, and Gln. The α and β structures, and the so-called "random" or unordered chain, appear to play different roles in the construction of the globular protein. If the α and β structures are thought of as the core or framework for a sculpture, then the extended chain is the clay in which the fine details of the work are executed. Lysozyme and carboxypeptidase are particularly good examples of this. Seemingly, even in individual enzyme molecules the α helix and β sheet play structural roles (see Chapter 2).

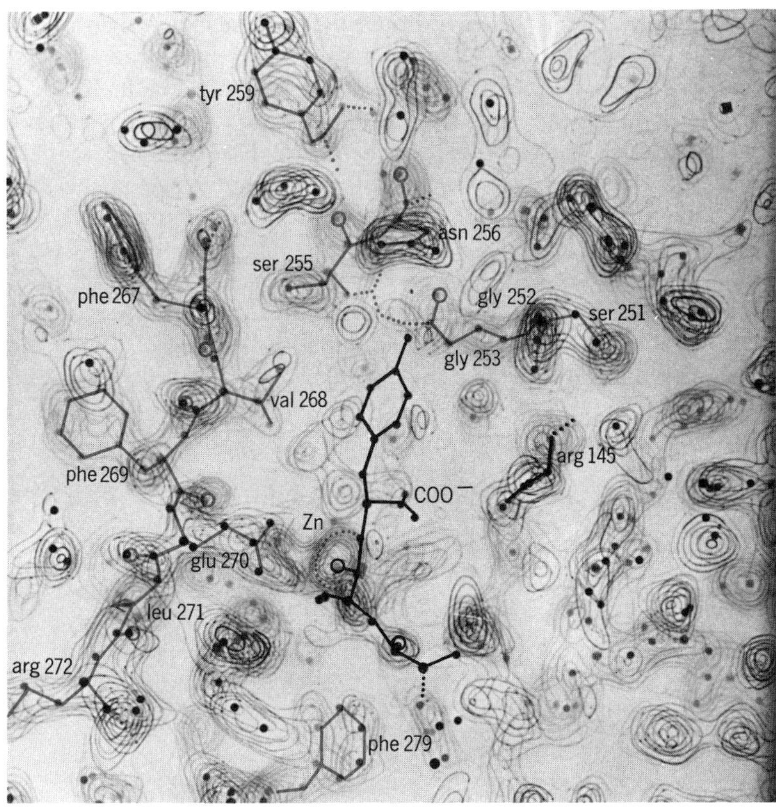

The five sections just above the ones shown on page 94. The carbobenzoxy substrate, shown in black, has been drawn in from a knowledge of the Gly-Tyr inhibitor position in other maps. β strand θ (residues 267–272) runs down the left side. The extended chain that defines the top of the hydrophobic pocket drops into these sections for residues 251–253, and then begins helix G with residues 255, 256, and 259 visible here. The zinc atom (Z) is enclosed within a dotted oval.

The basic driving force in keeping a globular protein folded is entropy, through the tendency of hydrophobic groups to segregate away from the aqueous environment. Conversely, many charged groups are found distributed over the surface of the molecule with the apparent function of making sure that these parts of the molecule remain outside. Hydrogen bonds are important within a helix or a sheet, but apparently exert little more than a directing influence in the folding of the helix or sheet substructures into a globular molecule.

All the enzymes described here have been catenases, so the discussion of principles is necessarily biased. But there seem to be three basic models of enzyme: those whose active site is in a crevice, a shallow depression, or a pit. The crevice enzymes—lysozyme, ribonuclease, and papain—act upon chains in which the bond to be severed is relatively exposed. One can imagine both the polyribonucleotide and the polysaccharide (even with the peptide cross-links) fitting into the crevice like wire into the jaws of cutting pliers. (Until it is learned what good papain does for the papaya plant, there is no point in speculating as to the form of the polypeptide chain which it has been evolved to cut.) The polypeptide digestive enzymes—chymotrypsin and carboxypeptidase—are designed differently. Most fibrous proteins, such as muscle myosin, keratin, and collagen, have an extensive secondary structure. Packed sheets, coiled coils, and supercables are the rule, and the exposed strand of polypeptide chain is rare. Under these circumstances, an enzyme that required a strand to fit into a deep crevice would be ineffective. Chymotrypsin appears to be designed to abut against a larger structure and to cut away at peptide bonds rather than to enfold a single strand. Carboxypeptidase has the logical form for an enzyme that cuts off the end of a chain; its active site is a pit into which the end can fit.

The specificity of lysozyme arises because of the inability of alternative binding sites in the crevice to accommodate the side chain of NAM. That of carboxypeptidase is brought about by the presence of a hydrophobic cavity at the active site, into which the

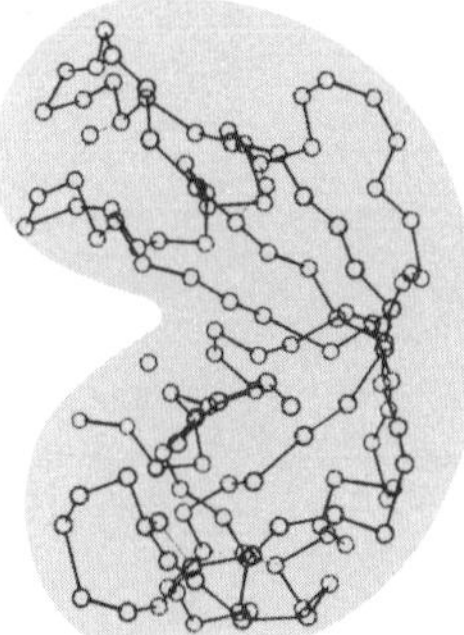

Ribonuclease A

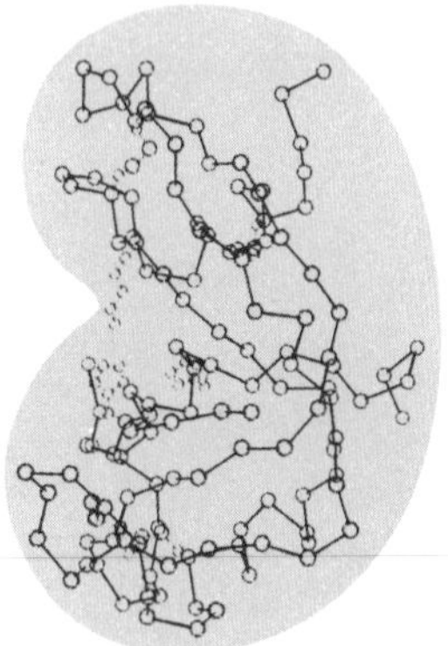

Ribonuclease S

Globular proteins drawn to a uniform scale

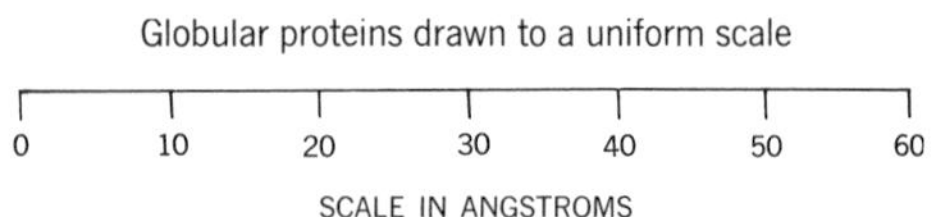

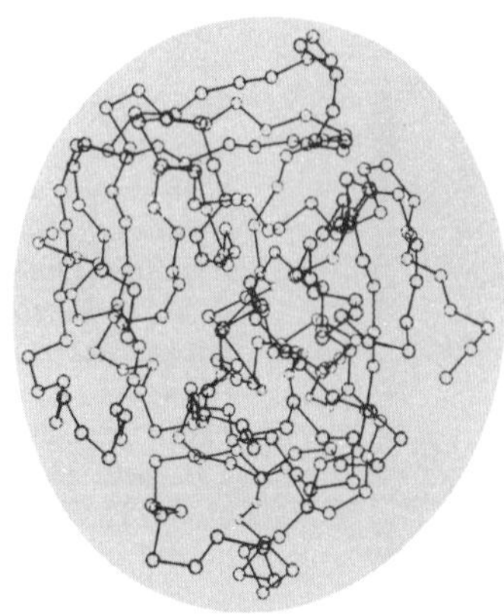

Chymotrypsin

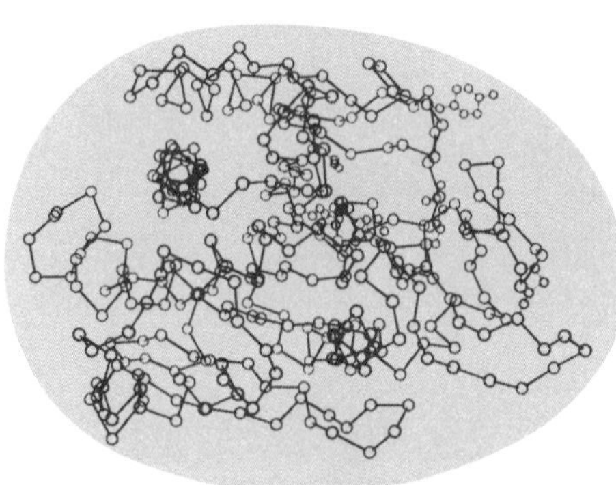

Carboxypeptidase A

side chain of the carboxyl terminal residue is fitted. Chymotrypsin has revealed a similar hydrophobic pocket in its active site depression in which the aromatic side group of the chain being cut is bound, and one would expect that trypsin would have an acidic group in a similar place. Ignoring the four Asp that are removed in the activation process, trypsin has 4 other Asp and 2 Glu. But Asp 72 and 153 and Glu 78 are far removed from the active site and are identical in chymotrypsin. Asp 102, by analogy with chymotrypsin, is presumably occupied in catalysis. Asp 194, although located in the active site, is the same in both enzymes. Glu 185 is the most likely candidate for the residue which interacts with a basic side group in trypsin and confers upon it its specificity (47). It is not only located very near His 57 and Ser 195 but is also part of a two-residue insertion that is absent in chymotrypsin.

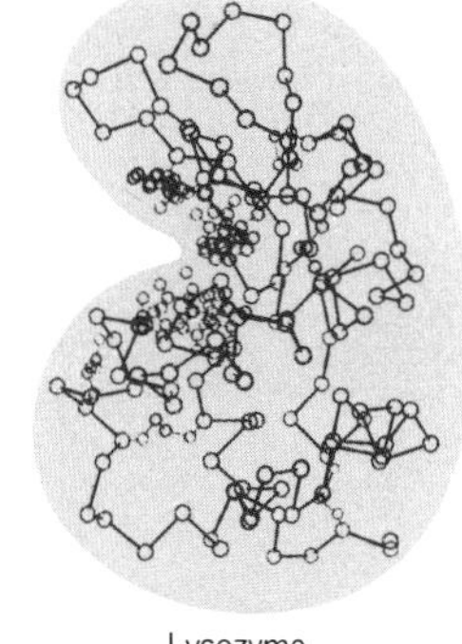
Lysozyme

It is idle to speculate about the geometry of active sites in the great number of enzymes that act on smaller molecules rather than chains, other than to predict that the types of steric specificity devices seen here will probably be generally valid. Human carbonic anhydrase (37), which catalyzes the reaction

$$H_2O + CO_2 \rightleftarrows H_2CO_3$$

has an active site rather like that of carboxypeptidase, a deep depression with a catalytically essential Zn atom at the bottom. It will be very interesting to have the structures of enzymes such as adenylate kinase, a small protein of molecular weight 21,000, which catalyzes the reaction

$$\mathrm{ATP + AMP = 2ADP}$$

and which is extremely sensitive to minor alterations in the substituents of the rings in the bound nucleotides. But most enzymes are far larger than the ones described here and involve the interaction of several subunits, some of which have catalytic sites and others which appear to exert control functions. This brings us to the last topic of the book, the organization of separate polypeptide chains into large-scale operating systems.

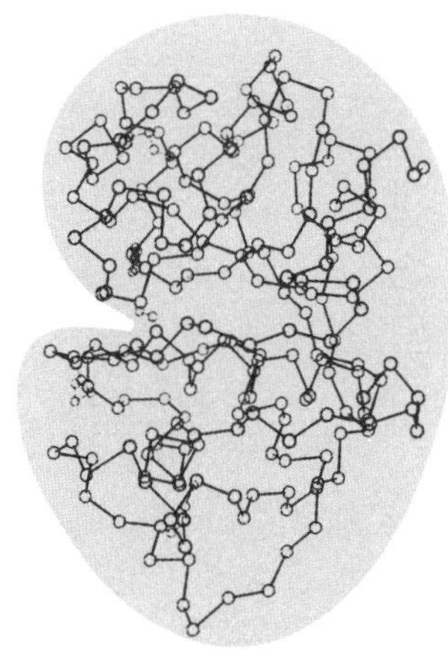
Papain

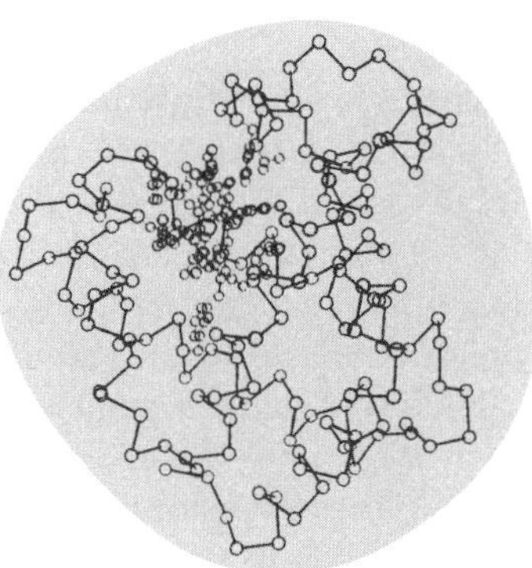
Myoglobin

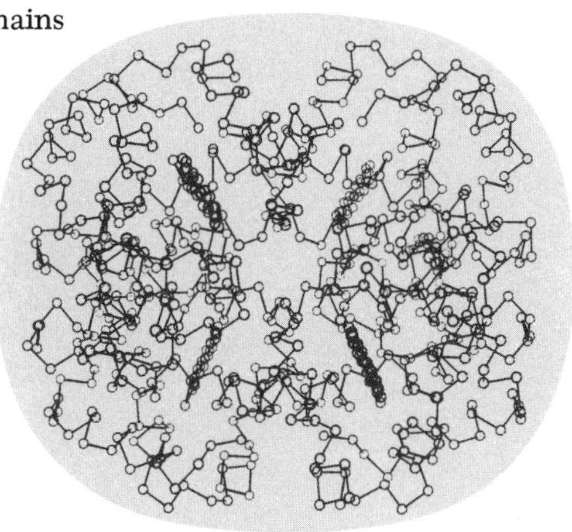
Hemoglobin

CHAPTER FIVE
THE NEXT STEP UP

5.1 WHAT SORT OF STEP?

No molecule by itself is alive, not even the overworked DNA molecule. "Life" is a behavior pattern of organized chemical systems of the proper degree and kind of complexity. An individual molecule of such a system can no more be alive than an individual bolt, wheel, or aileron in a jet airliner can fly. To say that DNA or any other molecule has the potential capacity for life in the right setting is as true, and nearly as misleading, as to say that the wings, fuel tank, and landing lights of a jet have the potential capacity for flight in the right setting. Moreover, even given the total collection of components in either of the systems we have been comparing, it is equally essential that these components be organized in precisely the right manner in order that the system be able to operate properly. If the much-abused doctrine of emergent properties has any validity, it is this: Complex systems can show properties that their individual components do not possess, properties that arise out of the way in which the components are organized. As we gain experience with computers and automatic control systems, it becomes more and more surprising that so simple a statement could have been dismissed on one hand as a platitude and damned on the other as the first step toward vitalism.

When we look to see what function the enzymes that we have been examining play in a living system, we often find that the sensible unit of study is not the individual folded polypeptide chain but a collection of such chains, in a setting organized both in space (structure) and in time (metabolism). This can be as simple as the cooperation between subunits in an enzyme molecule, or as complex as the organization of dehydrogenases, cytochromes, and ATP-synthesizing enzymes in the walls of a mitochondrion. In this last chapter we shall look at some examples of the organization of proteins into larger assemblies to try to see why such organization is advantageous and what new functions such assemblies can have.

5.2 PROTEINS WITH SUBUNITS

The enzymes that have been examined so far all have molecular weights of less than 30,000. But some of the most interesting enzymes are far larger, with molecular weights of several million. Where information is available, these have always been found to be built up from subunits, each a polypeptide chain of approximately 12,000–80,000. The table on page 101 summarizes the state of our knowledge as of 1967. Many of the proteins listed as having subunits over 100,000 may eventually prove to be made up of smaller units. Aspartate transcarbamylase, for example, is now known to have six catalytic subunits of 33,000 each and six regulatory subunits of 17,000 each, instead of the two and four chains of the table (1–3). Aldolase is also now known to have four chains and not three, two of one kind and two of another (4,5). To this table should be added mitochondrial malate dehydrogenase, with two chains of 32,500 each (6).

The dehydrogenases form an illustrative series of subunit enzymes, most of them being built up from two or four subunits of weight 30–45,000 (7). They are part of the machinery by which foodstuffs are broken down in a series of small, manageable steps, and the liberated energy is stored as adenosine triphosphate, ATP, until needed. A dehydrogenase takes a metabolite one small step along the way to complete oxidation by removing two hydrogen atoms and transferring them to a common carrier molecule such as nicotine adenine dinucleotide, NAD^+. The type reaction is

$$\text{metabolite}-H_2 + NAD^+ \rightleftharpoons \text{metabolite} + NADH + H^+$$

The reduced carrier molecules from many sources then feed into the cytochrome-containing terminal oxidation chain, where they are reoxidized for another round and where the free energy is used to synthesize ATP. Such general-purpose, continuously regenerated carrier molecules are called coenzymes. The loci of all of this activity in organisms with nuclear cells are the mitochondria.

Four dehydrogenases have been purified, crystallized, and given at least preliminary x-ray examinations, and there is the hope of some day having as detailed knowledge of these as for the smaller enzymes of Chapter 4. These are glyceraldehyde-3-phosphate dehydrogenase, GPDH (8); liver alcohol dehydrogenase, LADH (9, 10); malate dehydrogenase, MDH (11); and lactic dehydrogenase, LDH (12–14). GPDH is part of the process that prepares compounds such as glucose for entry into the citric acid cycle. It oxidizes and phosphorylates D-glyceraldehyde-3-phosphate to 1,3-diphospho D-glycerate and reduces NAD^+ to NADH at the same time. MDH oxidizes malate to oxaloacetate within the citric acid cycle, LADH oxidizes simple alcohols to aldehydes or ketones, and LDH reduces

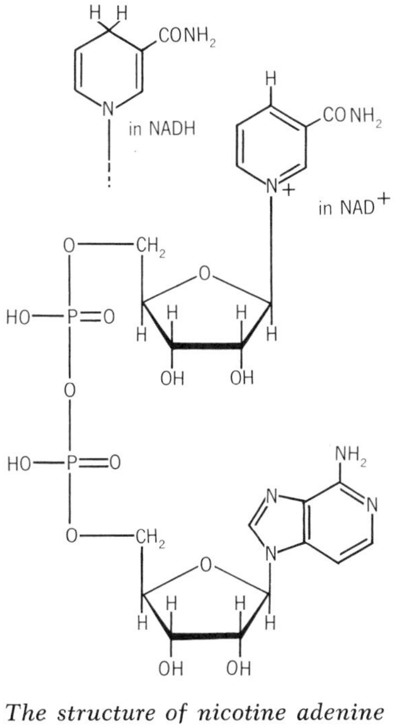

The structure of nicotine adenine dinucleotide, NAD^+, with the reduced configuration of the nicotine ring in NADH shown above. NAD^+ is called diphosopho-pyridine nucleotide, DPN^+, or Coenzyme 1 in the older literature.

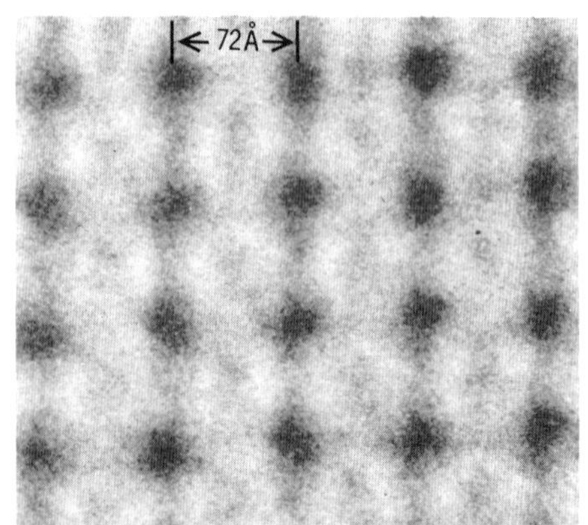

Electron micrograph of a crystal of LDH, showing individual molecules. The view is that indicated by an arrow in the bottom drawing. From (14).

LACTIC DEHYDROGENASE

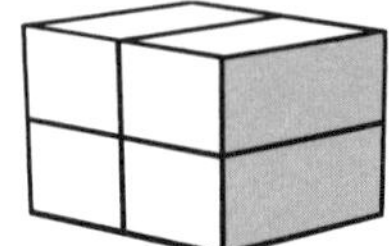

One molecule

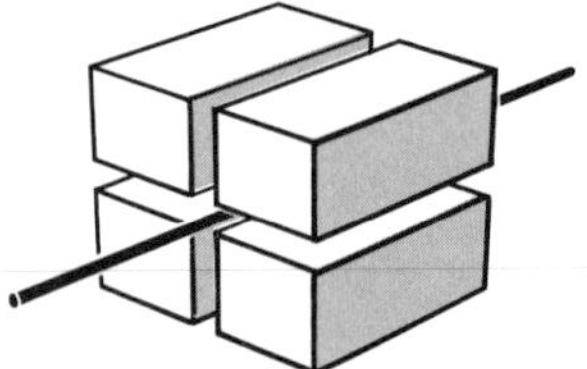

Exploded view of four subunits

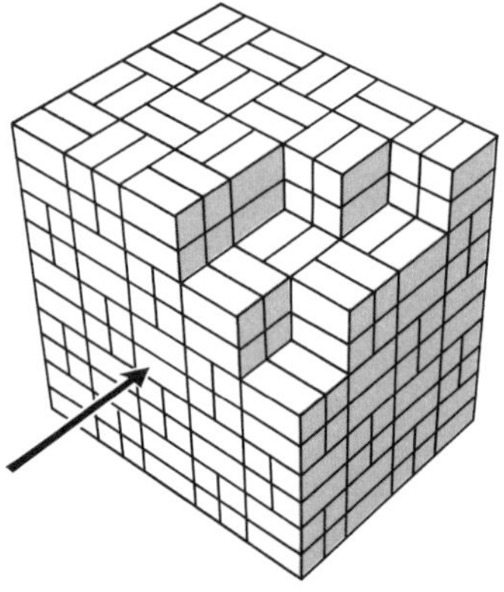

The packing of tetrameric molecules of LDH in the crystal. The arrow shows the view in the electron micrograph.

pyruvate to lactate in muscle just before the citric acid cycle if adequate oxygen to carry out the terminal oxidation process is lacking:

$$\underset{\text{pyruvate}}{CH_3{-}CO{-}COOH} + NADH + H^+ \rightleftharpoons \underset{\text{lactate}}{CH_3{-}CH(OH){-}COOH} + NAD^+$$

GPDH and LDH are both tetramers of 35,000-weight subunits. MDH has two subunits of this size, and LADH has two of 42,000. In these and similar enzymes with identical subunits, it is usual to find one active site per subunit. Then why the dimers and tetramers? What advantage is there in grouping monomers into larger structures?

There may be a conformational change, or at least a rearrangement of subunits, in these enzymes when NAD^+ or NADH is bound. The order of binding of coenzyme and substrate is fixed in many of these enzymes. In the pyruvate to lactate reaction of LDH, NADH must bind before pyruvate can do so, and lactate comes off the enzyme before NAD^+ (15,16). The binding of NADH or a large enough fragment of it causes a change in the way in which the molecule packs in the crystal, a change which suggests that the subunits have moved relative to one another by about 5 Å (13). MDH has the same sort of ordered binding (6), and the balance of the evidence suggests that LADH does also (17). The crystal form of LADH also changes when NADH is bound, and it may be significant that both ends of the long adenine–ribose–phosphate–phosphate–ribose–nicotine molecule of NADH are required to produce the change (10). GPDH not only needs NADH to bind substrate well, it is unstable in the absence of NADH or NAD^+ (20). GPDH also illustrates the fact that four chemically identical chains (18,19) do not necessarily behave identically in the molecule. The rate of binding and acylation of a substrate analogue to GPDH rises linearly with the amount of NAD^+ bound, up to two NAD^+ per molecule (20). More NAD^+ has only a minor effect. In addition, only *two* molecules of substrate analogue can be bound at one time to the enzyme. Although the polypeptide-chain characterization of GPDH is α_4, its functional description is more nearly $\alpha_2\beta_2$.

One advantage of an ordered binding pattern is the efficient use of coenzyme. The coenzyme will be generally in shorter supply than substrates such as pyruvate or lactate and would impede the reaction if allowed to get away. In the presence of an excess of pyruvate, the equilibrium will be shifted strongly in favor of binding. The pyruvate cannot bind in the absence of NADH, so the coenzyme is locked into place. When an excess of pyruvate exists, NADH is forced to spend most of its time helping the enzyme get rid of it. In circumstances where another metabolite has accumulated in excess, the NADH will be shunted to its enzyme instead. The overall result is a self-damping system of many enzymes, competing for NADH and mutually maintaining moderate concentrations of all the various metabolites in the energy-using chain.

Protein	Molecular weight	Subunits No.	Subunits Molecular weight
Insulin	11,466	2	5,733
Thrombin	31,000	(~3)	(~10,000)
β-Lactoglobulin	35,000	2	17,500
Bovine growth hormone	48,000	2	25,000
Avidin	68,300	4	18,000
Neurospora malate dehydrogenase	54,000	4	13,500
Hemoglobin	64,500	4	16,000
Tropomyosin	(75,000)	2	(35,000)
Glycerol-1-phosphate dehydrogenase	78,000	2	40,000
Alkaline phosphatase	80,000	2	40,000
Creatine kinase	80,000	2	40,000
Enolase	82,000	2	41,000
Liver alcohol dehydrogenase	84,000	2	42,000
Procarboxypeptidase	87,000	1	34,500
		2	25,000
Firefly luciferase	92,000	2	52,000
Hexokinase	96,000	4	24,000
Hemerythrin	107,000	8	13,500
Tryptophan synthetase A	29,000	1	29,000
Tryptophan synthetase B	117,000	2	60,000
Mammary glucose 6-phosphate dehydrogenase	130,000	2	63,000
Glyceraldehyde-3-phosphate dehydrogenase	140,000	4	37,000
Aldolase	142,000	3	50,000
Lactic dehydrogenase	150,000	4	35,000
Yeast alcohol dehydrogenase	150,000	4	37,000
Ceruloplasmin	151,000	8	18,000
Threonine deaminase	160,000	4	40,000
Cystathionine γ-synthetase	160,000	4	40,000
Thetin homocysteine methylpherase	180,000	3–4	50,000
Fumarase	194,000	4	48,500
Serum lipoprotein	200,000	6	36,500
Tryptophanase	220,000	2	(125,000)
Pyruvate kinase	237,000	4	57,200
Catalase	250,000	4	60,000
Acetoacetate decarboxylase	260,000	8	30,000
Mitochondrial adenosine triphosphatase	284,000	10	26,000
Aspartyl transcarbamylase	310,000	2	96,000
		4	30,000
Phosphoenolpyruvate carboxytransphosphorylase	430,000	(3–4)	(120,000)
Apoferritin	480,000	20	24,000
Urease	483,000	6	83,000
Phosphorylase	495,000	4	125,000
β-Galactosidase	520,000	4	130,000
	130,000	3–4	(40,000)
Myosin	620,000	3	200,000
Pyruvate carboxylase	660,000	4	165,000
	165,000	4	45,000
Thyroglobulin	669,000	2	335,000
Propionyl carboxylase	700,000	4	175,000
Lipoic reductase-transacetylase	1,600,000	60	27,000
Glutamic dehydrogenase	2,000,000	8	250,000
	250,000	5	50,000
Chlorocruorin	2,750,000	12	250,000
Turnip yellow mosaic virus	5,000,000	150	21,000
Poliomyelitis virus	5,500,000	130	27,000
Bushy stunt virus	9,000,000	120	60,000
Potato virus X	35,000,000	650	52,000
Tobacco mosaic virus	40,000,000	2130	17,500

PROTEINS WITH SUBUNITS

This table has been adapted and slightly abbreviated from one by I. M. Klotz in Handbook of Biochemistry, Selected Data for Molecular Biology, *H. A. Sober, ed., Chemical Rubber Co., Cleveland, 1968, p. C-47, and in* Science 155, *697 (1967). It contains the proteins known to have subunits held together by noncovalent bonds. Literature references for individual proteins will be found in the original table.*

PYRUVATE DEHYDROGENASE

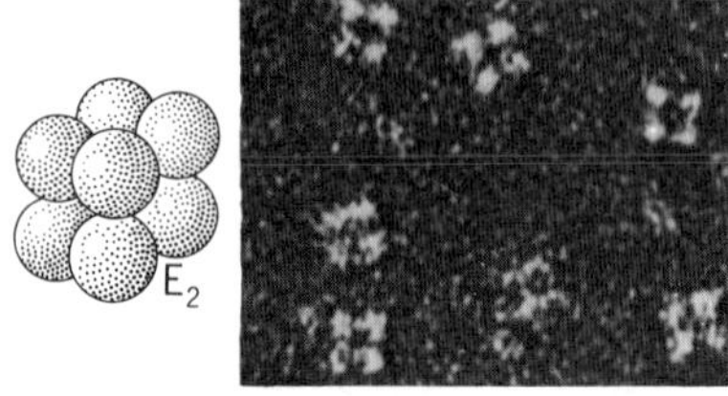

The E_2 cube of eight trimers. Each stippled ball is 3 subunits.

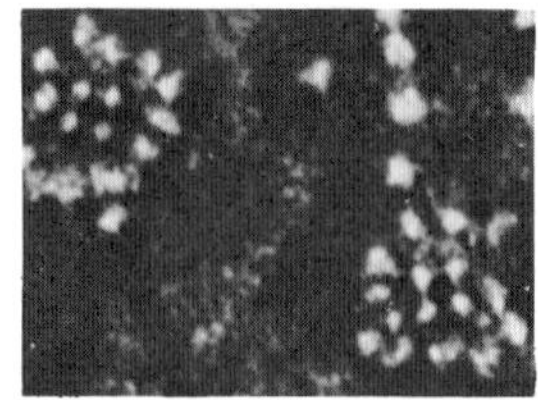

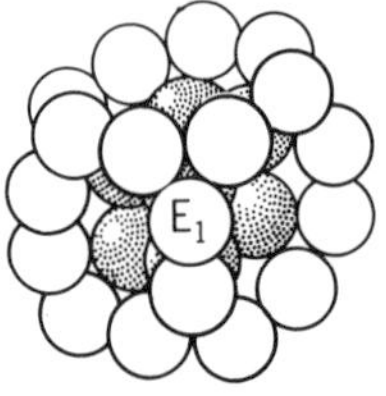

The E_2 cube with 24 molecules of E_1 (white balls) on the cube edges.

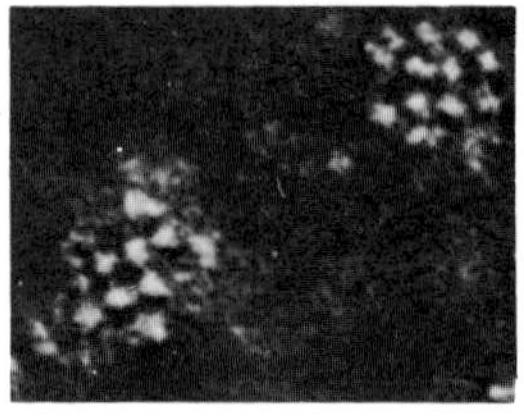

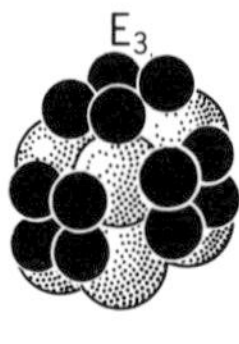

The E_2 cube with 24 molecules of E_3 (black balls) on the cube faces.

Organization of the total complex: E_1, E_2 (inside), and E_3.

The subunits in a complex not only need not be identical; they need not even be participating in the same enzymatic reaction. There are organized multienzyme complexes in which the products of one step become the substrates for the next and where it is advantageous to have the enzymes in close and ordered array. One such system is the pyruvate dehydrogenase complex (21). The overall reaction is

$$\underset{\text{pyruvate}}{CH_3-CO-COOH} + \underset{\text{coenzyme A}}{HS-CoA} + NAD^+ \rightleftharpoons$$

$$\rightleftharpoons \underset{\text{acetyl coenzyme A}}{CH_3-CO-S-CoA} + CO_2 + NADH + H^+$$

This is the final preparatory step before the entry of a metabolite into the citric acid cycle as acetyl coenzyme A. But the reaction actually occurs in three steps. In the first, the pyruvate binds to a thiamine pyrophosphate group that is attached to the first enzyme, E_1:

$$CH_3-CO-COOH + TPP-E_1 \rightleftharpoons CH_3-CH(OH)-TPP-E_1 + CO_2$$

In the second step, the just-formed α-hydroxyethyl group is transferred from the TPP on enzyme 1 to a long, free-swinging lipoate group that is bound to a lysine on enzyme 2. The lipoate is partially reduced and the α-hydroxyethyl oxidized to an acetyl group.

$$CH_3-CH(OH)-TPP-E_1 + \overset{S\text{———————}S}{\overset{|\qquad\qquad\quad|}{CH_2-CH_2-CH}}-R-E_2 \rightleftharpoons$$

$$\rightleftharpoons TPP-E_1 + \overset{SH\qquad\qquad S-CO-CH_3}{\overset{|\qquad\qquad\quad|}{CH_2-CH_2-CH}}-R-E_2$$

The symbol $-R-$ represents the remainder of the lipoyl molecule and the side chain of the lysine to which it is attached:

$$-CH_2-CH_2-CH_2-CH_2-CO-NH-CH_2-CH_2-CH_2-CH_2-C\alpha H\langle$$

In the third step, the acetyl group is transferred to coenzyme A and the lipoate is completely reduced:

$$HS-CoA + \overset{SH\qquad\qquad S-CO-CH_3}{\overset{|\qquad\qquad\quad|}{CH_2-CH_2-CH}}-R-E_2 \rightleftharpoons$$

$$\rightleftharpoons CH_3-CO-S-CoA + \overset{SH\qquad\qquad SH}{\overset{|\qquad\qquad\quad|}{CH_2-CH_2-CH}}-R-E_2$$

The acetyl CoA goes its way into the citric acid cycle, but the machinery must be recycled. The lipoate on enzyme 2 is reoxidized and the disulfide bond re-formed at the expense of reducing a flavin on enyzme 3; the flavin is then restored by reducing a NAD^+.

Space requirements are severe. The lipoate must be able to swing close enough to E_1 for the transfer of the α-hydroxyethyl group but also close enough to E_3 to be "recharged" at the end. Somewhere in the complex there must be a binding site for coenzyme A, and it, too, must be within the swinging range of the lipoate group.

The picture of the pyruvate dehydrogenase complex that has arisen from chemistry and from electron microscopy is shown on page 102. The total molecular weight of the complete complex is 4,440,-000. The framework of the complex is a cube, each of the eight corners of which contains three E_2 chains of 40,000 each. The six faces of the cube are each covered by four E_3 units of 55,000 each, and the 12 edges each have two E_1 units of 90,000 each. In one last measure of complexity, each of these E_1 units is actually an αβ dimer of two different polypeptide chains of similar size.

The E_2 enzyme is not only the framework; it is the organizer of the complex. It will form the cubic framework spontaneously and will then add the units of E_1 and E_3. The latter enzymes are known to be bound to E_2 and not to one another. And finally, one trimer of E_2—one corner of the cube—can add E_1 and E_3 to form a functioning subcomplex one eighth the size of the full assemblage.

5·3 ALLOSTERY AND FEEDBACK CONTROL

As biochemists have worked out the details of metabolism and synthesis over the past decade and a half, some interesting patterns of control have come to light. *Escherichia coli* bacteria and similar microorganisms have been the most productive, because their nutritional intake can be controlled, their life cycle is short, and enzyme-deficient mutants can be found that are unable to synthesize different essential compounds. Assume, for the moment, that compounds A, B, C, . . ., K, L, and M are the successive intermediates in a synthesis of an essential compound N. If a mutant lacks the enzyme for converting F to G, F will accumulate in excess. But it is often found that an artificial oversupply of the distant end product, N, will suppress a very early step in the chain, say the conversion of B to C, and will prevent an accumulation of the blocked intermediate, F (22). This control of an early step in a process by the end product is known in engineering as feedback control and is a mechanism for keeping a system in a steady state.

Many end products have been found which inhibit early steps in their synthesis: Arg, His, Ilu, Leu, Lys, Ser, Thr, Trp, Val, CTP and ATP, to list only a few. These control networks can be quite complex. Asp, for example, is a starting point in *E. coli* for the synthesis of Lys, Met, Ilu, and Thr (23), as shown to the right. The first step, the conversion of Asp to aspartyl phosphate, is catalyzed

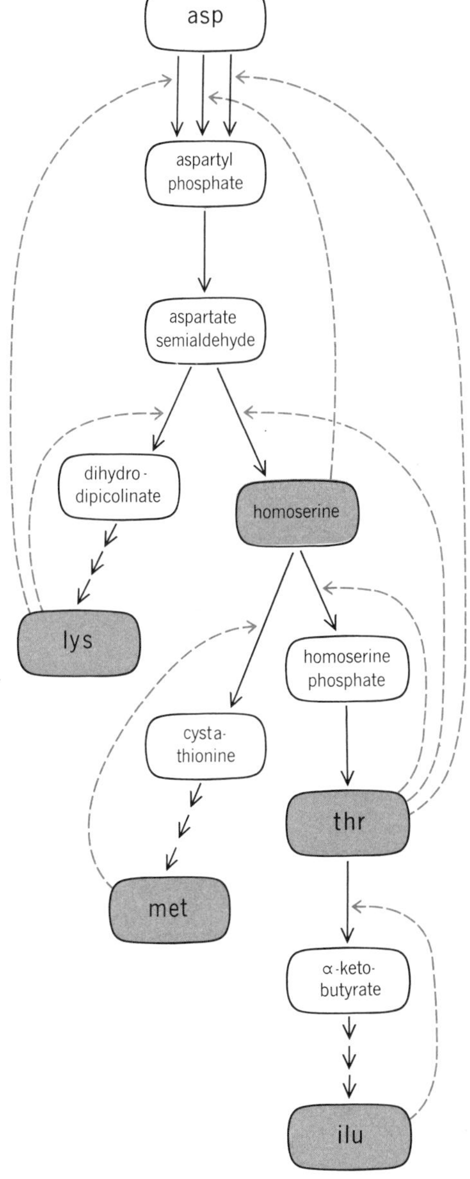

Asp is the starting point for the synthesis of Lys, Met, Ilu, and Thr in E. coli *bacteria. Many products inhibit intermediate steps of synthesis and provide feedback control.*
⟶ *Synthetic step*
→→ *Unspecified number of intermediate synthetic steps*
- - → *Feedback inhibition of enzyme Products that cause feedback control are in color.*

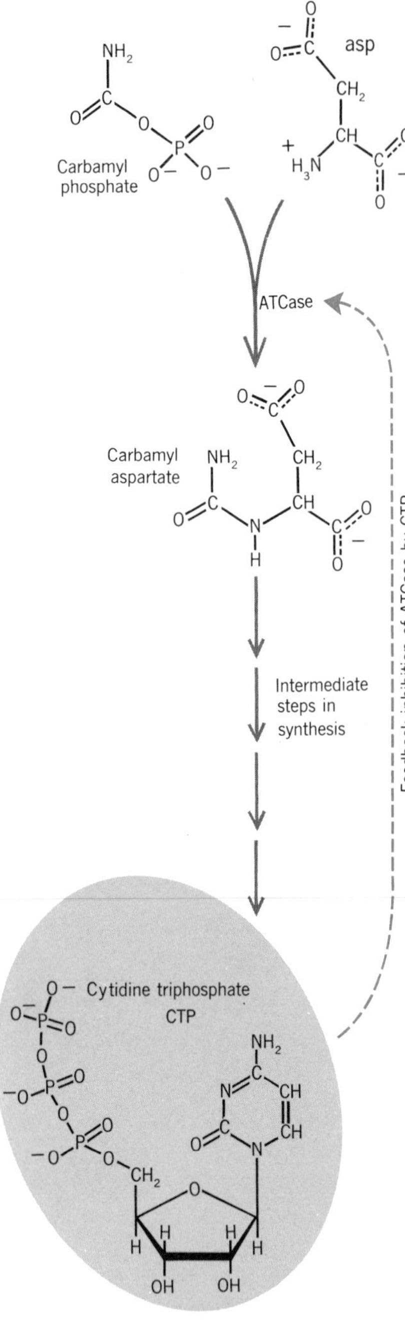

Aspartate transcarbamylase catalyzes the first of a series of steps in the production of CTP, which in turn is an inhibitor of aspartate transcarbamylase.

by aspartokinase, an enzyme that exists in three molecular forms. One form is inhibited by Lys, one by Thr, and one by homoserine, a synthetic intermediate. An oversupply of any one of the three inhibitors leads to a partial but not a complete suppression of all synthesis.

But there is an additional degree of sophistication, in the form of "gating" control just beyond each of the main branch points of synthesis. An excess of Lys inhibits the enzymatic conversion of aspartate semialdehyde to dihydrodipicolinate and encourages the production of homoserine instead. Homoserine itself is shunted into the synthesis of either Met or Thr by an excess of the *other* component. Even Ilu, the most distant of the end products, slows down its own synthesis from Thr when present in large amounts. Thr controls its own production by inhibiting key enzymes at no less than three different points. There appears to be one logical flaw in the scheme shown on page 103—an excess of Thr will also shut down the production of Met. And sure enough, *E. coli* requires an outside supply of Met for proper growth in the presence of large amounts of Thr in the growth medium.

The usefulness of such feedback control in maintaining steady conditions in a living organism is obvious, and many examples are known in carbohydrate and nucleic-acid metabolism as well (23, 24). The champion feedback control enzyme of all appears to be *E. coli* glutamine synthetase, a complex of 12 identical subunits of 48,000 each, which can be inhibited by *eight* different end products from the manifold pathways of glutamine metabolism (25–27).

The first detailed study of this feedback control was made on aspartate transcarbamylase, ATCase. This is the first enzyme in the conversion of Asp to cytidine triphosphate, CTP (left), and is inhibited by an excess of CTP. It is difficult to imagine competitive binding at the active site, for CTP bears no obvious structural similarity with the true substrates. Moreover, Gerhart and Pardee (28) found that:

1. The binding of Asp to ATCase shows the sigmoid curve which is the sign of cooperative interaction between subunits (recall hemoglobin).
2. Although the enzyme is inhibited by CTP, the cooperativity of the active sites remains.
3. CTP acts by diminishing the binding constant of Asp to its active site, but does not affect the rate of catalysis of Asp once it is bound.
4. CTP binds to a site quite distinct from the catalytic site. In fact, the two kinds of site, catalytic and regulatory, occur on entirely different subunits. After a period of confusion (1), it now ap-

pears that ATCase has six catalytic and six regulatory subunits, of weights 33,000 and 17,000 each.

Out of the work on hemoglobin and ATCase came the best theoretical explanation of these phenomena, Monod's *allostery* theory.

Monod and coworkers at the Institut Pasteur have proposed a model system for such multiunit control enzymes (29–31). The modification of an enzymic reaction by a compound of quite different shape from the true substrate, by virtue of binding to the enzyme at some place other than the active site, has been named *allostery* ("different structure"). In this model, the binding of this *allosteric effector* at its site alters the conformation of the subunits in unison and thereby alters the enzymatic properties of all the active sites on the molecule in the same way.

Hemoglobin can be considered an "honorary enzyme" in this theory—a special case in which the heme is both allosteric site and ligand-binding site and the oxygen molecule is both allosteric effector and binding ligand (30,32). Each subunit is assumed to have two conformations, R (oxy) and T (deoxy). The O_2 is assumed to have appreciable affinity only for the R conformation, and all the subunits are assumed to be in the same state, changing from one to the other in concert. The equilibria involved in O_2 binding in this model are shown to the right. With a value of the conformational equilibrium constant of 9000, the expression for fractional heme binding, Y, fits the observed hemoglobin saturation curves (page 45) fully as well as the purely empirical Hill equation at the bottom of page 45. As Koshland has pointed out (33), this does not prove that the model is right, only that it is not obviously wrong. Nevertheless, its success is encouraging.

This allosteric model would propose that the equilibrium point for unoxygenated hemoglobin is overwhelmingly in favor of the open deoxy form ($L = 9000$), and that only the binding of O_2 makes the compact oxyhemoglobin conformation preferable. The first O_2 binds with difficulty, having to fight this lopsided equilibrium and change the conformation of the protein. Later O_2 molecules bind with greater ease, being successively more insulated from the conformation-changing step.

In the more general case, allosteric *inhibitors* or *activators*, binding at sites other than the active site, can affect the binding of substrates by shifting the equilibrium between conformations. The more complex equilibrium equations do succeed in accounting for the kinetic data from many enzymes. The model is criticized most severely for its assumption of symmetry and of the simultaneous conformational change of all subunits (33–35). But the criticisms imply that the model should be made more sophisticated, not that it be abandoned.

$$
\begin{array}{cc}
T_0 & \\
\Updownarrow & \frac{T_0}{R_0} = L \\
R_0 & \\
-F \Updownarrow +F & \frac{4R_0 \cdot F}{R_1} = K \\
R_1 & \\
-F \Updownarrow +F & \frac{3R_1 \cdot F}{2R_2} = K \\
R_2 & \\
-F \Updownarrow +F & \frac{2R_2 \cdot F}{3R_3} = K \\
R_3 & \\
-F \Updownarrow +F & \frac{R_3 \cdot F}{4R_4} = K \\
R_4 &
\end{array}
$$

EQUILIBRIA IN THE ALLOSTERIC MODEL FOR HEMOGLOBIN

T_o = *Concentration of protein in deoxy configuration*

R_n = *Concentration of protein in oxy configuration with n bound oxygen molecules*

F = *Concentration of unbound oxygen molecules*

L = *Conformational equilibrium constant*

K = *Microscopic dissociation constant for binding of O_2 to heme in the oxy configuration* = F × (unbound sites)/(bound sites)

N_o = *Total sites occupied by ligand* = $R_1 + 2R_2 + 3R_3 + 4R_4$

N_a = *Total sites available* = $4T_o + 4R_o + 4R_1 + 4R_2 + 4R_3 + 4R_4$

Y = *Fractional occupancy* = N_o/N_a

$$Y = \frac{\alpha(1+\alpha)^3}{(1+\alpha)^4 + L} \quad \textit{where } \alpha = F/K$$

5.4 GAMMA GLOBULINS*

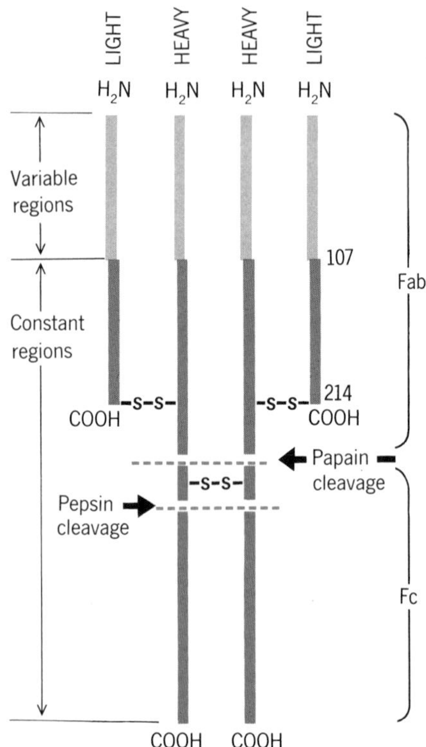

The chains in a gamma globulin molecule, their cross-linking, and the cleavage points with two enzymes. Variable parts of light and heavy chains are shown in color.

Among the most impressive of all the working multichain protein molecules are the gamma globulin antibodies. In higher vertebrates there are three main classes, γG, γA, and γM. The most common, γG, is made up of molecules of weight 150,000 daltons, each with two "heavy chains" of 52,000 and two "light chains" of 23,000. These chains are connected by disulfide bridges, as shown to the left. γA molecules are similar in size and construction, and γM molecules appear to be built from five such units.

The gamma globulins are important because of their ability to récognize foreign materials in the blood serum or body secretions, to combine with them, and to remove them by precipitation. They work best against alien proteins or carbohydrates, the materials from which the walls of invading microorganisms would be built. G molecules have been shown to be bivalent; each molecule has two active sites that can bind with the foreign material to build up a three-dimensional network and produce an insoluble aggregate (lower left). The two sites on any one antibody molecule are identical, and antibody sites are both incredibly specific and diverse. Experiments with haptens, small artificial analogues of antigens, have shown that antibody active sites are as sensitive to minor changes in the molecules they recognize as are the active sites of enzymes. On the other hand, the number of different molecules that can be recognized by one or another of the many antibodies that we produce runs into the thousands. Any one foreign cell or microorganism will be "recognized" by many strains of antibody molecule because of different chemical groups on its surface, and the precipitate from such an antibody response will be made up of a great many different antibodies. Antigen precipitation, therefore, is not a useful way to obtain pure antibody. But it happens that humans and certain strains of laboratory mice are susceptible to a cancerous growth known as a myeloma tumor, in which one gamma globulin–producing plasma cell propagates wildly and produces a mass of tumor cells, all of which are secreting an identical globulin. In some cases the whole gamma globulin molecule accumulates in the serum; in others, only the light chain is produced and is eliminated in the urine as "Bence–Jones protein." Although activity against specific antigens has been identified in only a few of these myeloma globulins, the preparations are pure proteins of one amino acid sequence, and all evidence suggests that they are perfectly good antibody molecules. Both the amino acid sequence work and preliminary x-ray studies have been made on myeloma proteins.

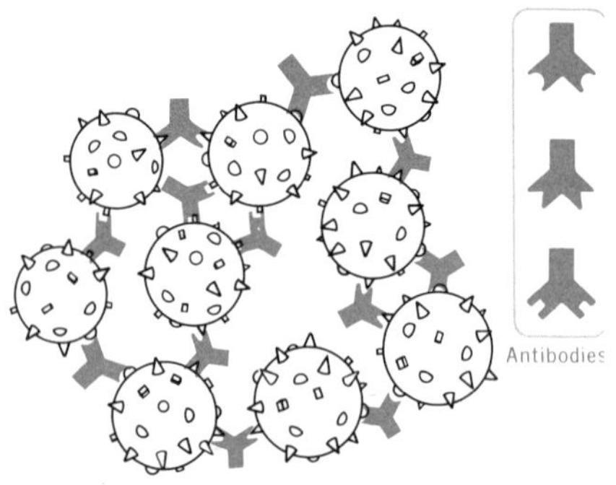

The cross-linking of small virus particles with antibodies using three different antigenic sites on the virus protein. Compare the top right of the opposite page.

* A good introduction to antibody structure can be found in (36) and a more comprehensive treatment in (37).

The sequence work on the light and heavy chains has built up a puzzling story. Both chains are involved in the active recognition sites, in each case in the first 100 residues of the chain. It was once thought that the total number of antigens that could be recognized by antibodies was far too large for a separate antibody molecule to be encoded in the genetic material of the host for each one. There was assumed to be some general-purpose antibody that molded itself around an antigen upon first exposure and then made copies of itself in some way to produce more molecules, ready to recognize more of the same intruder. A better theory proposes that a limited number of basic antibody molecule sequences is coded in the genetic material of the host and that these genes somehow mutate during cell division to lead to the thousands of different antibody molecules that are observed.

The pattern of variation of amino acid sequences in the myeloma proteins is curious. Virtually all the variation in sequence, in both heavy and light chains, occurs in the first 100 residues of the chain, those which are believed to be responsible for the specificity of the antibody. Because the second half of the light chains is so invariant, and because it has at least been shown that the constant halves of all light chains are coded for by a common gene, it has been proposed that the two halves arise from different segments of DNA. In any one plasma cell that is to produce gamma globulin, the "constant-half" DNA is associated during cell division with one of the many "variable-half" lengths of DNA before protein production begins. The template theory is known to be incorrect, but a decision cannot yet be made between the idea of mutation of genes during cell division and of combination of pre-evolved genes for heavy and light halves.

Whatever the mechanism of production of this variability, it is very old. Sequence comparisons have shown that there are often greater similarities in the variable regions of the light chains of mice, rabbits, and humans than there are within any one species, suggesting that the pattern of inherited variation of sequences arose before the time of divergence of these mammalian lines.

It is obviously of great interest to find out how such a vital molecule is put together—how its active sites are arranged and how the four chains are folded to form them. But x-ray analysis has been frustrated until recently by the difficulty of obtaining pure, homogeneous material and of crystallizing even the myeloma proteins. Papain cuts the molecule at the place shown on page 106 to produce an Fc ("crystalline") fragment containing the stumps of two heavy chains and two Fab ("antigen-binding") fragments, each a light chain and the remaining half of a heavy chain. The Fc fragment was easiest to crystallize, lacking as it does the variable-sequence parts of the chains (38,39). More recently, the Fab fragment (40) and then the entire gamma globulin molecule from a human myeloma

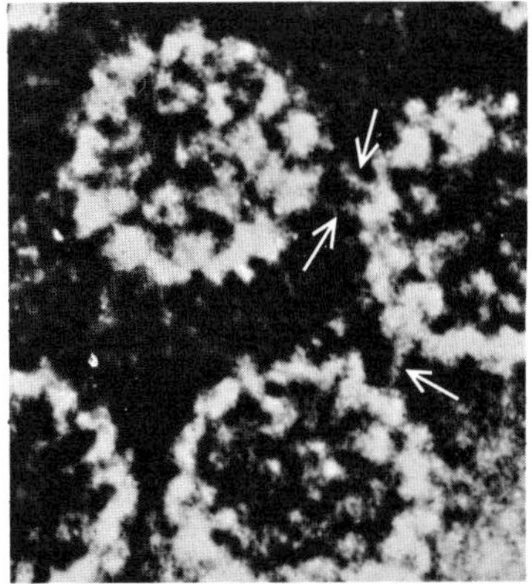

Polyoma virus particles (radius 450 Å) crosslinked by γ G antibody *molecules. From (42).*

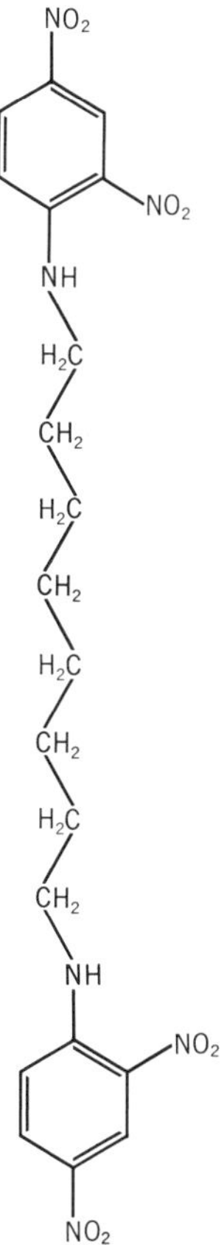

The bifunctional hapten which was used for the electron microscopy of rings of γ G antibody molecules shown on the next page.

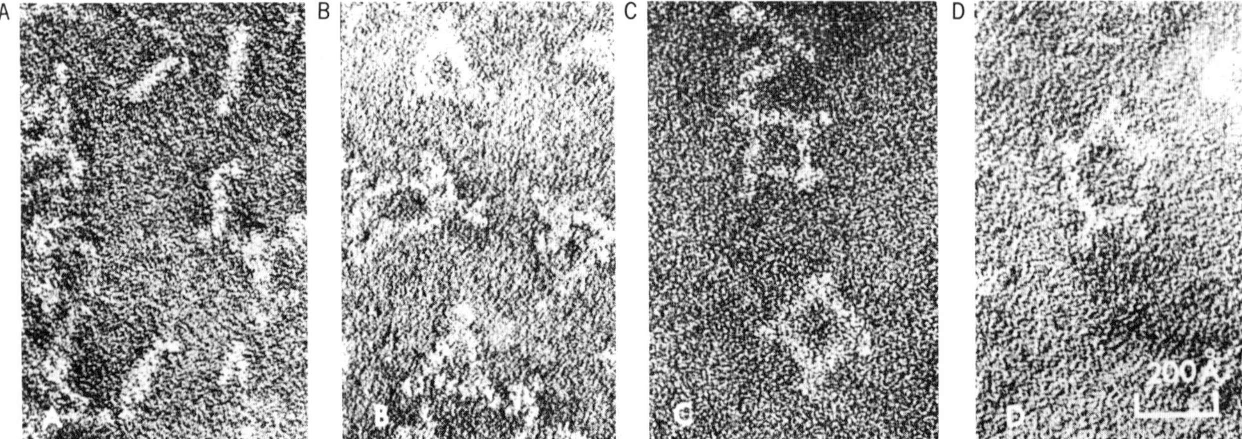

PHOTOGRAPH OF ANTIBODY MOLECULES

Polymeric rings of γ G gamma globulin antibodies, linked at their active binding sites by the divalent hapten on the preceding page. A, dimers; B, trimers; C, tetramers; D, pentamers. From (45).

tumor strain (41) have been crystallized, and x-ray structure analysis is now possible. Both the Fc fragment and the whole molecule have been shown from preliminary x-ray examination to possess a twofold axis of symmetry.

The electron microscope has been more informative so far. Antibodies have been seen cross-linking virus particles (42) (page 107) and the smaller molecules of ferritin (43). But in what surely must be one of the triumphs of electron microscopy, the late R. C. Valentine succeeded in obtaining photographs of antibody bound to haptens so small that they disappeared from view altogether (44, 45). The hapten used (bottom, page 107) was a polymethylenediamine chain with an immunologically active dinitrophenyl group at each end. With eight or more carbon atoms in the central chain, the two ends could interact with different antibody molecules, coupling them by their active sites. Such random coupling could produce dimers and rings of three, four, five, and more antibody molecules, such as are seen above. Digestion with pepsin, which is known to remove the Fc part of the molecule without breaking the disulfide bridge that connects the two Fab parts, left the rings intact but eliminated the knobs that occurred at the vertices (below).

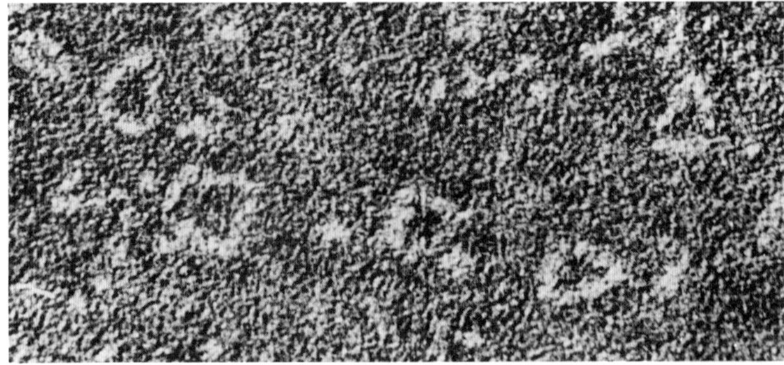

The same antibody rings, after digestion with pepsin to remove the Fc fragments at the vertices.

The picture of the trimer and of the gamma globulin molecule that emerged is shown to the right. The molecule is shaped like a capital Y, the stem being the Fc part and the two arms the Fab segments. Each arm must contain one light chain and one half of a heavy chain and must have an active site at its upper end. The angle between the arms is observed to vary, so the arms must be "hinged" to the stem by a polypeptide chain somewhere near the middle of the heavy chains.

The gamma globulin antibody is the most unsymmetrical protein macromolecule that we have encountered so far and scarcely deserves the designation "globular" protein. Yet it is shaped well for the job it performs; a hinged molecule with variable distance between its two active sites would be more efficient at cross-linking foreign particles than would a rigid rod with a fixed geometry. The next step is to find out the detailed structure of the binding sites of a pure antibody strain. But it is highly unlikely that there will be any real surprises, and the sites will probably turn out to be much like those involved in enzyme specificity.

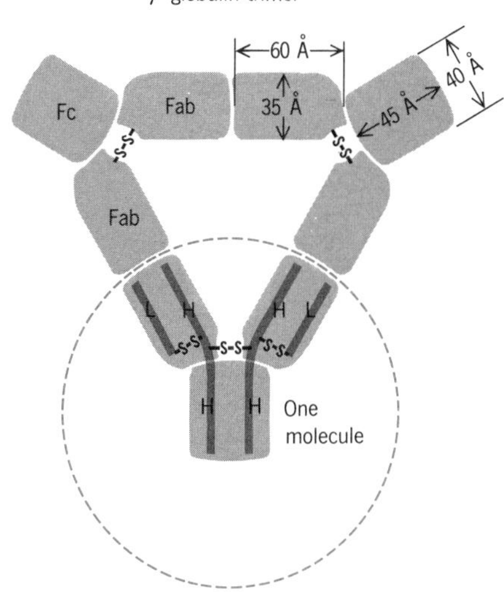

DIAGRAM OF ANTIBODY MOLECULES

The gamma globulin antibody trimer, as seen in B on the opposite page. The probable locations of the heavy (H) and light (L) chains are shown in one of the three molecules.

5.5 THE SERUM COMPLEMENT SYSTEM*

The most complex macromolecular system we shall consider, the serum complement system, shows many of the traits that have been discussed so far: antigen recognition, enzyme activation, subunit interaction, ordered binding, organization in time as well as in space, and allosteric control. It also introduces a new factor, amplification.

The precipitation of foreign bodies by agglutination is only one response of antibodies. With large objects such as bacteria, a more efficient defense is to kill them by punching holes in them and letting their contents drain out. This is one major purpose of the complement system. This system is made up of nine different proteins, present in the serum in varying amounts (table, page 110). As its name suggests, it is a system that completes the action begun by the antigen–antibody binding, by interacting with bound antibody and making it even more destructive to the intruder cell. With a cast of characters of over a dozen, the easiest way to keep the complicated set of events straight is to think of them as a drama in five acts.

Act I in the process is the familiar one in which a host antibody recognizes a chemical group on the surface of an alien microorganism and binds to it. There will be many different sites, and many different antibodies will be able to bind. But this mech-

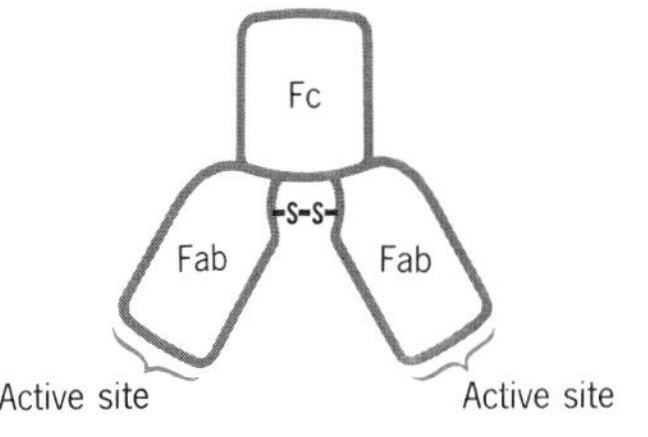

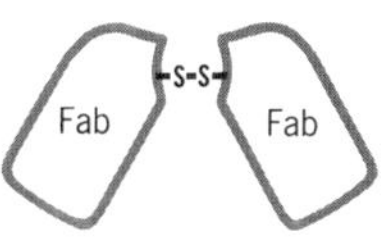

One whole γ globulin molecule (above left). Molecule after digestion of Fc portion with pepsin (above right).

* This section is largely based on a recent review by Müller-Eberhard (46). More background information is also available in (47).

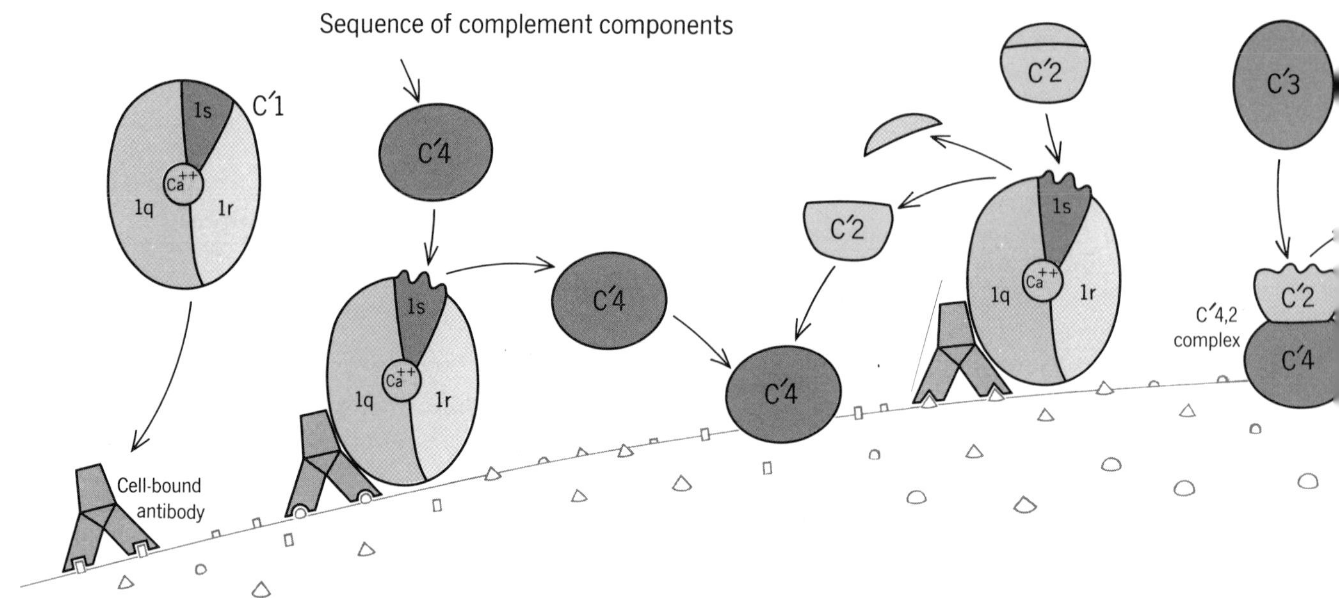

Component	Molecular weight	Concentration in serum (Mg/ml)
1q	400,000	100—200
1r	(160,000)	—
1r	(80,000)	22
2	115,000	10
3	240,000	1200
4	230,000	430
5	(210,000)	75
6	(120,000)	—
7	(120,000)	—
8	(190,000)	—
9	79,000	1—2

The component of the human complement system. Values in parentheses are only rough estimates from sedimentation constants.

anism alone would be like tying elephants together with thread. Something more lethal is needed.

In Act II, the machinery of the plot is developed. The first serum complement component, C′1, combines with the bound antibodies and becomes an enzyme. Component C′1 is actually three subunits, 1q, 1r, and 1s, held together by a calcium ion. (If the calcium ion is tied up by a reagent such as EDTA, the three subunits fall apart.) Subunit 1s is a proenzyme, which is converted to an active enzyme when the entire complex binds to antibody. Yet it is *subunit 1q* and not 1s that binds. Hence some form of allosteric interaction must exist between subunits, so that binding to 1q opens up the active site on 1s. Neither 1q nor 1r has any detectable enzymatic activity of its own. But activated C′1 (that is, 1s) appears to be an esterase. It is inhibited by DFP like trypsin and several other proteases and esterases.

The purpose of this enzymatic site on the surface of the alien cell is to prepare the next serum component, C′4, for binding to the cell membrane. This 230,000-weight molecule, diffusing to the cell from the serum, is activated by enzyme C′1 so that it can attach to other proteins, apparently by hydrophobic interactions. It may be that a piece is removed, uncovering part of the hydrophobic interior, or that C′4 is partially unfolded. If it collides with the foreign cell wall immediately thereafter, it sticks. If it does not collide quickly, it loses its binding ability and remains in the serum as inactivated component. An exposed hydrophobic surface is entropically unfavorable, so it may be that if the surface is not covered by binding to another protein rapidly, the polypeptide chains spontaneously refold to bury the hydrophobic groups. This is an example of a time-dependent process and is a safety device to ensure that these trigger

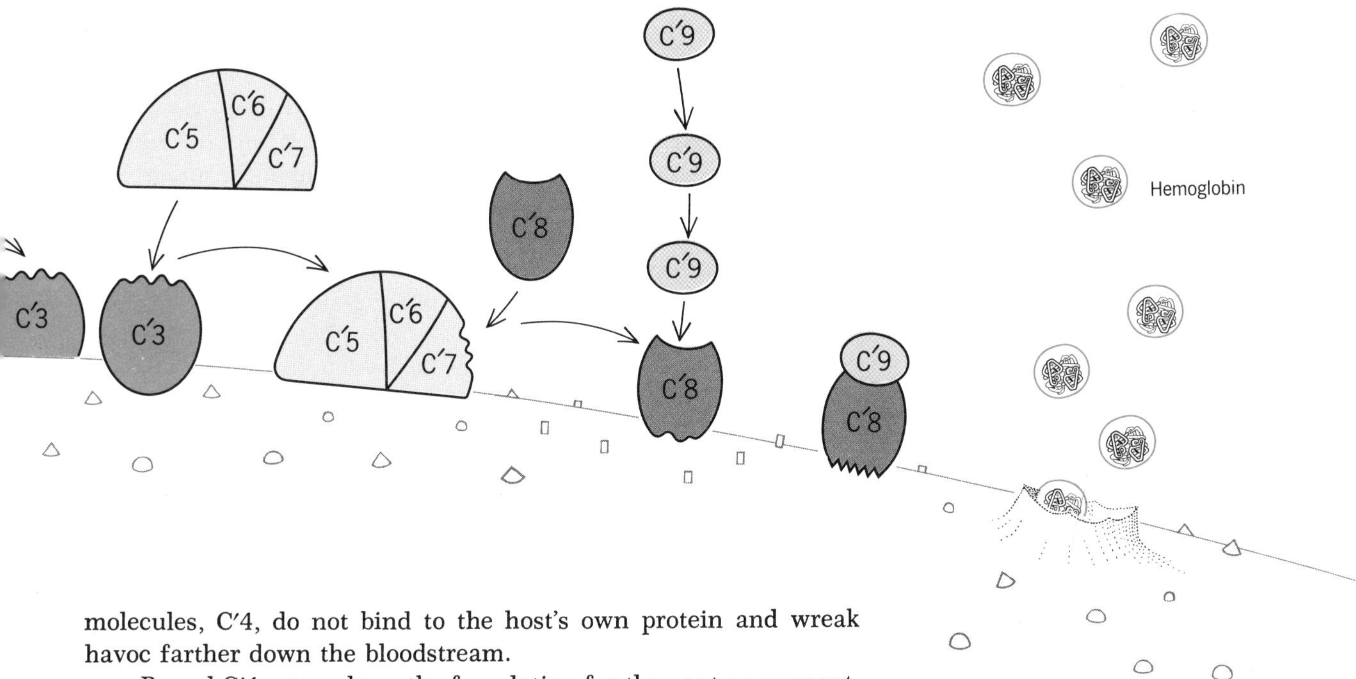

molecules, C′4, do not bind to the host's own protein and wreak havoc farther down the bloodstream.

The lysis of a red blood cell by the serum complement system.

Bound C′4 acts only as the foundation for the next component. A molecule of weight 115,000, C′2, is split into two parts by enzyme C′1, one of 76,000 and one of 39,000. The smaller fragment is lost, but the larger piece attaches to C′4 to form a second enzymatically active unit. These new enzyme centers, of which there may be as many as 1000 on one alien cell, are the scenes of the next activity.

Act II, as is often the case, was involved, intricate, and inconclusive. But things begin to develop again in Act III. Each of the new enzymatic C′4, 2 centers now activates from two or three to perhaps 100 molecules of component C′3, which bind all around on the neighboring cell surface. This is the amplification step, by which the effect of each original antibody molecule is multiplied many times. As before, there is a safety device. If activated C′3 does not bind to the surface immediately, it goes dead and is carried in the serum as inert protein. The enzymatic site built up from a molecule of C′4 and one of C′2 has been named "C′3 convertase," which is an elegant way of saying that the chemists haven't the foggiest notion of how it works. Component C′3, once properly bound around the C′4, 2 site, is a peptidase enzyme.

Act IV is less well understood. Components C′5, C′6, and C′7, either sequentially or as a unit, become activated and bind to the cell surface. The enzymatic properties of both C′4, 2 and C′3 are needed for this step. The Sparafucile of our drama then appears in the form of component C′8, which is acted upon by C′7 and then binds directly to the cell membrane. At this point the membrane first appears to be damaged, for if the cell is a red blood cell, slow leakage of hemoglobin molecules is observed.

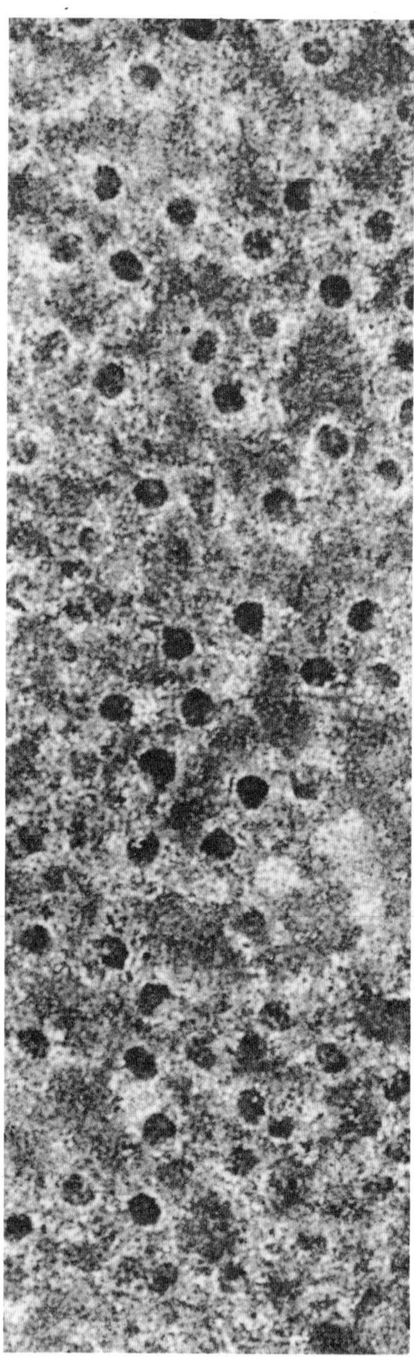

A fragment of a red cell membrane that has been sensitized with antibody and lysed with human serum complement. The holes are approximately 100 Å in diameter. From (48).

The *coup de grâce* occurs in Act V. The last component, C′9, binds to C′8 (and not to the cell wall itself) and induces C′8 to produce holes that are visible for the first time in the electron microscope (left) and that cause the breakdown and death of the intruder. This again must be an example of allostery, with the binding of C′9 to one part of the C′8 molecule affecting its interactions with the membrane at another site. The effect of C′9 can be mimicked by phenanthrolene.

The overall result is that a relatively few antibodies bound to a cell membrane can produce a great many punctures. Furthermore, the complement machinery is general. Antibodies for many different combining sites can all participate in the binding of C′1 and the subsequent steps. In at least the C′3 step and perhaps in others, the destructive power of the system is amplified. But for the mechanism to work, the components must be present at the right place and *at the right time.*

The mechanism is complex, because the membrane that it is designed to attack is itself complex. One can imagine, at some much earlier point, a simpler membrane and a simple attack mechanism. But as the membrane evolved for protection, so did the destructive system, in a reciprocal escalation of selection pressures, until there evolved the elaborate mechanism that we see today.

The serum complement system is one example of multienzyme machinery that illustrates many of the ideas of earlier sections. Others could have been chosen, such as the actin-myosin mechanism of muscle (49,50), the organization of enzymes for energy storage in the mitochondria (51,52), the assembly of polypeptide chains at the ribsomes (53), or the infection and lysis of a bacterium by a virus (54). It must not be forgotten that enzymes such as we have been studying, although formidably complicated at first, are really only the gears and camshafts of the important piece of machinery, the functioning organism. But for the first time since enzymes were discovered in the nineteenth century, we can begin to specify how these gears and camshafts fit together to carry out the tasks for which they are so essential.

SUGGESTIONS FOR FURTHER READING*

CHAPTER 1

The choice of chemical elements for life:

L. J. Henderson, *The Fitness of the Environment,* Beacon Press, Boston, 1913, 1958—A classic work with an amazing amount of relevance and chemical insight even now.

The applicability of the laws of physics and chemistry to living systems:

E. Schrödinger, *What is Life?* Cambridge Univ. Press, New York, 1944, 1962—Some pointed questions by a quantum physicist, which helped to touch off the explosive growth of molecular biology.

H. Blum, *Time's Arrow and Evolution,* Princeton Univ. Press, Princeton, N.J., 1969—First published in 1951, but still one of the best treatments of the thermodynamics of living systems and of the role of entropy.

H. J. Morowitz, *Energy Flow in Biology,* Academic Press, New York, 1958, and B. C. Goodwin, *Temporal Organization in Cells,* Academic Press, New York, 1963—Two attempts to build a statistical mechanics for living systems. They build upon the ideas of order and entropy that Blum started.

The origin of life:

A. I. Oparin, *Life, Its Nature, Origin and Development,* Oliver & Boyd, London, 1961—A clear summary by a pioneer who has been making people think about such questions for over forty years.

A. Dauvillier, *The Photochemical Origin of Life,* Academic Press, New York, 1965—Much more assertive than Oparin, and a book to be read with caution. Much useful information.

J. D. Bernal, *The Origin of Life,* World Publishing, Cleveland, 1967—An excellent introduction by another of the pioneers in the field.

The nucleic acid code and the chemistry of heredity:

J. D. Watson, *Molecular Biology of the Gene,* Benjamin, New York, 1965—An authoritative and readable book written with zest by one of the men who started it all.

J. D. Watson, *The Double Helix,* Atheneum, New York, 1968—His version of how it all started.

F. H. C. Crick, "The Genetic Code: III," *Sci. American* (October 1966) p. 55.—A good summary of the current status of the nucleic acid code, when supplemented by references (19) or (20).

Resonance theory and chemical bonding:

L. Pauling, *The Nature of the Chemical Bond,* 3rd ed., Cornell Univ. Press, Ithaca, N.Y., 1960—A standard text, and a classic, but one that gives molecular orbital theory short shrift.

---

* The best starting point for information on recent protein structure research is now the *Cold Spring Harbor Symposium on Quantitative Biology,* Vol. 36 (1971). At this conference on structure and function in proteins, almost every currently active protein structure group was represented, as well as nuclear magnetic resonance workers and enzyme chemists. *Annual Review of Biochemistry* volumes are also worth consulting for summaries of current work. A good introduction to the methods of protein crystallography is to be found in a chapter by David Eisenberg in Volume I of the 3rd Edition of *The Enzymes* (P. Boyer, Editor; Academic Press, New York, 1970).

—R. E. D., 1971

CHAPTER 2

The folding of polypeptide chains:

This field is too new, and most of the information must be dug out of the original literature. The review by Schellman and Schellman (2) is a good starting point, but their angle notations are now obsolete. See (3) and the footnote on page 24.

The structures of fibrous proteins:

R. Borasky, ed., *Ultrastructure of Protein Fibers,* Academic Press, New York, 1963—Very informative, but it assumes some knowledge of the field.

J. B. Finean, *Engström–Finean Biological Ultrastructure,* 2nd ed., Academic Press, New York, 1967—Excellent and well-illustrated textbook. Good starting point.

J. Gross, "Collagen," *Sci. American* (May 1961), p. 120.

K. Laki, "The Clotting of Fibrinogen," *Sci. American* (March 1962), p. 60.

K. R. Porter and C. Franzini-Armstrong, "The Sarcoplasmic Reticulum," *Sci. American* (March 1965), p. 73.

H. E. Huxley, "The Mechanism of Muscular Contraction," *Sci. American* (December 1965), p. 13.

CHAPTER 3

The structures of the heme proteins:

There is no substitute for the original literature. See especially the papers published in *Scientific American* and *Science* (the Nobel lectures) for an introduction.

The mitochondrion, the terminal oxidation chain, and cythochrome c:

A. Lehninger, *The Mitochondrion,* Benjamin, New York, 1964—A good starting point, at the senior or graduate level.

D. Keilin, *The History of Cell Respiration and Cytochrome,* Cambridge Univ. Press, New York, 1966—A very personal and absorbing book by the man who invented the word "cytochrome."

E. Racker, "The Membrane of the Mitochondrion," *Sci American* (February 1968), p. 32.

Molecular evolution:

Most of the literature on this subject is represented in references (26–33) and (39). The Bryson–Vogel symposium (28) and the entire Brookhaven symposium (39) are especially good.

CHAPTER 4

Molecular structure of enzymes:

Everything except the original literature is hopeless, and anything before 1960 is obsolete. In addition to the proteins mentioned in the chapter, the nonheme iron protein rubredoxin is being studied by L. Jensen at the University of Washington and subtilisin BPN by Joe Kraut at the University of California at San Diego. Most new results in this field tend to appear in *Nature, Proceedings of the National Academy of Science of the United States, Journal of Molecular Biology, Journal of Biological Chemistry,* and *Proceedings of the Royal Society* (London), Series B.

CHAPTER 5

The Ciba Foundation Symposium volume (54) and *Molecular Organization and Biological Function* (52) are both excellent. For an attempt to treat the problem of organization in more abstract terms, see J. M. Reiner, *The Organism as an Adaptive Control System,* Prentice-Hall, Englewood Cliffs, N. J., 1968.

REFERENCES

CHAPTER 1

1. L. Pauling, R. B. Corey, and H. R. Branson, *Proc. Natl. Acad. Sci. U.S.* 37, 205 (1951).
2. "Table of Interatomic Distances and Configurations in Molecules and Ions," *Special Publication No. 11*, The Chemical Society, Burlington House, London, 1958.
3. R. E. Marsh and J. Donohue, *Advan. Protein Chem.* 22, 235 (1967).
4. M. F. Perutz, *Nature* 167, 1053 (1951).
5. R. E. Dickerson, "*X-Ray Analysis and Protein structure*," in *The Proteins*, H. Neurath, ed., Vol. II, Academic Press, New York, 1964.
6. J. A. Schellman and C. Schellman, "Conformation of Polypeptide Chains in Proteins," in *The Proteins*, H. Neurath, ed., Vol. II, Academic Press, New York, 1964.
7. J. C. Kendrew, *Brookhaven Symp. Biol.* 15, 216 (1962).
8. M. F. Perutz, *J. Mol. Biol* 13, 646 (1965).
9. M. F. Perutz, J. C. Kendrew, and H. C. Watson, *J. Mol. Biol*, 13, 669 (1965).
10. C. C. F. Blake, G. A. Mair, A. C. T. North, D. C. Phillips, and V. R. Sarma, *Proc. Roy. Soc. (London)* B167, 365 (1967).
11. H. W. Wyckoff, K. D. Hardman, N. M. Allewell, T. Inagami, L. N. Johnson, and F. M. Richards, *J. Biol. Chem.* 242, 3984 (1967).
12. P. B. Sigler, D. M. Blow, B. W. Matthews, and R. Henderson, *J. Mol. Biol.* 35, 143 (1968).
13. W. N. Lipscomb, J. A. Hartsuck, G. N. Reeke, F. A. Quiocho, P. H. Bethge, M. L. Ludwig, T. A. Steitz, and P. Bethge, *Brookhaven Symp. Biol.* 21 (1968), 24.
14. H. C. Watson, *Progress in Stereochemistry*, Vol. 4, Butterworth, London, 1968.
15. C. Tanford, *J. Am. Chem. Soc.* 84, 4240 (1962).
16. C. J. Epstein, *Nature* 215, 355 (1967).
17. W. Kauzmann, *Advan. Protein Chem.* 14, 1 (1959).
18. E. L. Smith, in *Structure and Function of Cytochromes*, K. Okunuki, M. D. Kamen, and I. Sekuzu, eds., Univ. Tokyo Press, 1968.
19. "The Genetic Code," *Cold Spring Harbor Symp. Quant Biol.* 31 (1966).
20. D. Zipser, *Sci. American* (April 1968), p. 44.
21. G. A. Jeffrey and R. K. McMullan, *Progr. Inorg. Chem.* 8, 65 (1967).

CHAPTER 2

1. C. Ramakrishnan and G. N. Ramachandran, *Biophys. J.* 5, 909 (1965).
2. J. A. Schellman and C. Schellman, in *The Proteins*, H. Neurath, ed., Vol. II, Academic Press, New York, 1964.
3. J. T. Edsall, P. J. Flory, J. C. Kendrew, A. M. Liquori, G. Nemethy, G. N. Ramachandran, and H. A. Scheraga, *J. Mol. Biol.* 15, 399 (1966) and 20, 589 (1966).
4. S. Arnott and A. J. Wonacott, *J. Mol. Biol.* 21, 371 (1966).
5. G. R. Tristram and R. H. Smith, *Advan. Protein Chem.* 18, 227 (1963).
6. L. Pauling and R. B. Corey, *Proc. Natl. Acad. Sci. U.S.* 37, 729 (1951).
7. F. Lucas, J. T. B. Shaw, and S. G. Smith, *Biochem. J.* 66, 468 (1957).
8. R. E. Marsh, R. B. Corey, and L. Pauling, *Biochem. Biophys. Acta* 16, 1 (1955).
9. F. Lucas, J. T. B. Shaw, and S. G. Smith, *J. Textile Inst.* 46, T440 (1955).
10. R. E. Dickerson, in *The Proteins*, H. Neurath, ed., Vol. II, Academic Press, New York, 1964.
11. R. Borasky, ed., *Ultrastructure of Protein Fibers*, Academic Press, New York, 1963.
12. B. K. Filshie and G. E. Rogers, *J. Mol. Biol.* 3, 784 (1961).
13. G. E. Rogers, *J. Ultrastruct. Res.* 2, 309 (1959).
14. M. G. Dobb, F. R. Johnston, J. A. Nott, L. Oster, J. Sikorski, and W. S. Simpson, *J. Textile Inst.* 53, T153 (1961).
15. A. Rich and F. H. C. Crick, *J. Mol. Biol.* 3, 483 (1961).
16. G. N. Ramachandran, ed., in *Aspects of Protein Structure*, Academic Press, New York, 1963, p. 39.

CHAPTER 3

1. A. White, P. Handler, and E. L. Smith, *Principles of Biochemistry*, 3rd ed., McGraw-Hill, New York, 1964, ch. 34.
2. J. S. Fruton and S. Simmonds, *General Biochemistry*, 2nd ed., Wiley, New York, 1958, ch. 6.
3. H. Lehmann and R. G. Huntsman, *Man's Hae-*

moglobins, North-Holland, Amsterdam, 1966, ch. 2.
4. J. C. Kendrew, *Science 139,* 1259 (1963).
5. M. F. Perutz, *Science 140,* 863 (1963).
6. J. C. Kendrew, R. E. Dickerson, B. E. Strandberg, R. G. Hart, D. R. Davies, D. C. Phillips and V. C. Shore, *Nature 185,* 422 (1960).
7. J. C. Kendrew, *Sci. American* (December 1961), p. 96.
8. J. C. Kendrew, H. C. Watson, B. E. Strandberg, R. E. Dickerson, D. C. Phillips, and V. C. Shore, *Nature 190,* 665 (1961).
9. J. C. Kendrew, *Brookhaven Symp. Biol. 15,* 216 (1962).
10. H. C. Watson, *Progress in Stereochemistry,* Vol. 4, Butterworth, London, 1968.
11. M. O. Dayhoff and R. V. Eck, *Atlas of Protein Sequence and Structure, 1967–68,* National Biomedical Research Foundation, Silver Spring, Md.
12. R. A. Bradshaw and F. R. N. Gurd, *J. Biol. Chem. 244,* 2167 (1969).
13. M. F. Perutz, M. G. Rossmann, A. F. Cullis, H. Muirhead, G. Will, and A. C. T. North, *Nature 185,* 416 (1960).
14. A. F. Cullis, H. Muirhead, M. F. Perutz, M. G. Rossmann, and A. C. T. North, *Proc. Roy. Soc. (London) A265,* 15 (1961).
15. A. F. Cullis, H. Muirhead, M. F. Perutz, M. G. Rossmann, and A. C. T. North, *Proc. Roy. Soc. (London) A265,* 161 (1962).
16. M. F. Perutz, *Sci. American* (November 1964), p. 64.
17. M. F. Perutz, *J. Mol. Biol. 13,* 646 (1965).
18. M. F. Perutz, J. C. Kendrew, and H. C. Watson, *J. Mol. Biol. 13,* 669 (1965).
19. H. Muirhead, J. M. Cox, L. Mazzarella, and M. F. Perutz, *J. Mol. Biol. 28,* 117 (1967).
19a. W. Bolton, J. M. Cox, and M. F. Perutz, *J. Mol. Biol. 33,* 283 (1968).
20. M. F. Perutz, H. Muirhead, J. M. Cox, L. C. G. Goaman, F. S. Mathews, E. L. McGandy, and L. E. Webb, *Nature 219,* 29 (1968).
21. M. F. Perutz, H. Muirhead, J. M. Cox, and L. C. G. Goaman, *Nature 219,* 131 (1968).
22. M. F. Perutz and H. Lehmann, *Nature 219,* 902 (1968).
23. R. Huber, H. Formanek, and O. Epp, *Naturwiss. 55,* 75 (1968).
24. J. D. Bernal, I. Fankuchen, and M. F. Perutz, *Nature 141,* 523 (1938).
25. J. V. Dacie, N. K. Shinton, P. J. Gaffney, Jr., R. W. Carrell, and H. Lehmann, *Nature 216,* 663 (1967).
26. V. M. Ingram, *The Hemoglobins in Genetics and Evolution,* Columbia Univ. Press, New York, 1963.
27. E. Zuckerkandl, *Sci. American* (June 1965), p. 110.
28. V. Bryson and H. J. Vogel, eds., *Evolving Genes and Proteins,* Academic Press, New York, 1965.
29. G. H. Dixon, in *Essays in Biochemistry,* P. N. Campbell and G. D. Greville, eds., Vol. 2, Academic Press, New York, 1966, p. 147.
30. W. M. Fitch and E. Margoliash, *Science 155,* 279 (1967).
31. C. Nolan and E. Margoliash, *Ann. Rev. Biochem. 37,* 727 (1968).
32. E. Margoliash and W. M. Fitch, *Ann. N.Y. Acad. Sci. 151,* 359 (1968).
33. J. L. King and T. H. Jukes, *Science 164,* 788 (1969).
34. E. A. Padlan and W. E. Love, *Nature 220,* 376 (1968).
35. K. Ando, H. Matsubara, and K. Okunuki, *Proc. Japan Acad. 41,* 79 (1965).
36. R. E. Dickerson, M. L. Kopka, J. E. Weinzierl, J. C. Varnum, D. Eisenberg, and E. Margoliash, *J. Biol. Chem. 242,* 3015 (1967).
37. R. E. Dickerson, M. L. Kopka, J. E. Weinzierl, J. C. Varnum, D. Eisenberg, and E. Margoliash, in *Structure and Function of Cytochromes,* K. Okunuki, M. D. Kamen, and I. Sekuzu, eds., Univ. Tokyo Press, 1968.
38. E. Margoliash and A. Schejter, *Advan. Protein Chem. 21,* 113 (1966).
39. E. Margoliash, W. M. Fitch, and R. E. Dickerson, in *Structure, Function and Evolution in Proteins, Brookhaven Symp. Biol. 21,* 259 (1968).

CHAPTER 4

1. C. O'Sullivan and F. W. Tompson, *J. Chem. Soc. 57,* 834 (1890).
2. W. M. Bayliss, *The Nature of Enzyme Action,* 3rd ed., Longmans, London, 1914.
3. J. B. S. Haldane, *Enzymes,* Longmans, London, 1930.
4. J. D. Bernal and D. Crowfoot (D. C. Hodgkin), *Nature 133,* 794 (1934).
5. E. Fischer, *Ber. Deut. Chem. Ges. 27,* 2985 (1894).
6. R. E. Canfield and A. K. Liu, *J. Biol. Chem. 240,* 1997 (1965).
7. J. Jolles, J. Jauregui-Adell, I. Bernier, and P. Jolles, *Biochem. Biophys. Acta 78,* 668 (1963).
8. R. E. Canfield (1970), private communication.
9. K. Brew, T. C. Vanaman, and R. L. Hill, *J. Biol. Chem. 242,* 3747 (1967).
10. C. C. F. Blake, D. F. Koenig, G. A. Mair, A. C. T. North, D. C. Phillips, and V. R. Sarma, *Nature 206,* 757 (1965).
11. L. N. Johnson and D. C. Phillips, *Nature 206,* 759 (1965).
12. D. C. Phillips, *Sci. American* (November 1966), p. 78.
13. D. C. Phillips, *Proc. Natl. Acad. Sci. U.S. 57,* 484 (1967).
14. C. C. F. Blake, G. A. Mair, A. C. T. North, D. C. Phillips, and V. R. Sarma, *Proc. Roy. Soc. (London) B167,* 365 (1967).
15. C. C. F. Blake, L. N. Johnson, G. A. Mair,

A. C. T. North, D. C. Phillips, and V. R. Sarma, *Proc. Roy. Soc. (London) B167*, 378 (1967).
16. L. N. Johnson, *Proc. Roy. Soc. (London) B167*, 439 (1967).
17. H. C. Watson, *Progress in Stereochemistry*, Vol. 4, Butterworth, London, 1968.
18. R. U. Lemieux and G. Huber, *Can. J. Chem. 33*, 128 (1955).
19. R. U. Rupley and V. Gates, *Proc. Natl. Acad. Sci. U.S. 57*, 496 (1967).
20. F. W. Dahlquist, T. Rand-Meir, and M. A. Raftery, *Proc. Natl. Acad. Sci. U.S. 61*, 1194 (1968).
20a. T.-Y. Lin and D. E. Koshland, Jr., *J. Biol.Chem. 244*, 505 (1969).
20b. S. M. Parsons and M. A. Raftery, *Biochemistry 8*, 701 (1969).
21. K. Brew and P. N. Campbell, *Biochem. J. 102*, 258 (1967).
22. D. C. Phillips, *Proceedings of the Plenary Sessions, Seventh International Congress of Biochemistry*, Tokyo, 1967, p. 63.
23. K. Brew, T. C. Vanaman, and R. L. Hill, *Proc. Natl. Acad. Sci. U.S. 59*, 491 (1968).
24. G. Kartha, J. Bello, and D. Harker, *Nature 213*, 862 (1967).
25. H. W. Wyckoff, K. D. Hardman, N. M. Allewell, T. Inagami, D. Tsernoglou, L. N. Johnson, and F. M. Richards, *J. Biol. Chem. 242*, 3749 (1967).
26. H. W. Wyckoff, K. D. Hardman, N. M. Allewell, T. Inagami, L. N. Johnson, and F. M. Richards, *J. Biol. Chem. 242*, 3984 (1967).
26a. B. Gutte and R. B. Merrifield, *J. Am. Chem. Soc. 91*, 501 (1969); R. G. Denkewalter, D. F. Veber, F. W. Holly, and R. Hirschmann, *J. Am. Chem. Soc. 91*, 503 (1969).
27. B. W. Matthews, P. B. Sigler, R. Henderson, and D. M. Blow, *Nature 214*, 652 (1967).
28. P. B. Sigler, D. M. Blow, B. W. Matthews, and R. Henderson, *J. Mol. Biol. 35*, 143 (1968).
28a. D. M. Blow, J. J. Birktoft, and B. S. Hartley, *Nature 221*, 337 (1969).
28b. H. C. Watson and D. M. Shotton, private communication (1969).
28c. C. S. Wright, R. A. Alden, and J. A. Kraut, *Nature 221*, 235 (1969).
29. W. N. Lipscomb, J. C. Coppola, J. A. Hartsuck, M. L. Ludwig, H. Muirhead, J. Searl, and T. A. Steitz, *J. Mol. Biol. 19*, 423 (1966).
30. M. L. Ludwig, J. A. Hartsuck, T. A. Steitz, H. Muirhead, J. C. Coppola, G. N. Reeke, and W. N. Lipscomb, *Proc. Natl. Acad. Sci. U.S. 57*, 511 (1967).
31. T. A. Steitz, M. L. Ludwig, F. A. Quiocho, and W. N. Lipscomb, *J. Biol Chem. 242*, 4662 (1967).
32. G. N. Reeke, J. A. Hartsuck, M. L. Ludwig, F. A. Quiocho, T. A. Steitz, and W. N. Lipscomb, *Proc. Natl. Acad. Sci. U.S. 58*, 2220 (1967).
33. W. N. Lipscomb, J. A. Hartsuck, G. N. Reeke, F. A. Quiocho, P. H. Bethge, M. L. Ludwig, T. A. Steitz, H. Muirhead, and J. C. Coppola, *Brookhaven Symp. Biol. 21*, 24 (1968).
34. W. N. Lipscomb, G. N. Reeke, J. A. Hartsuck, F. A. Quiocho, and P. H. Bethge, *Proc. Roy. Soc.* (London) Series B (1969), in press.
35. J. Drenth, J. N. Jansonius, and B. G. Wolthers, *J. Mol. Biol. 24*, 449 (1967).
36. J. Drenth, J. N. Jansonius, R. Koekoek, H. M. Swen, and B. G. Wolthers, *Nature 218*, 929 (1968).
37. K. Fridborg, K. K. Kannan, A. Liljas, J. Lundin, B. Strandberg, R. Strandberg, B. Tilander, and G. Wiren, *J. Mol. Biol. 25*, 505 (1967).
38. D. H. Meadows and O. Jardetzky, *Proc. Natl. Acad. Sci. U.S. 61*, 406 (1968).
39. R. B. Corey, O. Battfay, D. A. Bruekner, and F. G. Mark, *Biochem. Biophys. Acta 94*, 535 (1965).
40. H. Neurath, *Federation Proc. 23*, 1 (1964).
41. K. A. Walsh and H. Neurath, *Proc. Natl. Acad. Sci. U.S. 52*, 884 (1964).
42. H. Neurath, *Sci. American* (December 1964), p. 68.
43. J. Kraut, H. T. Wright, M. Kellerman, and S. T. Freer, *Proc. Natl. Acad. Sci. U.S. 58*, 304 (1967).
44. A. Light, R. Frater, J. R. Kimmel, and E. L. Smith, *Proc. Natl. Acad. Sci. U.S. 52*, 1276 (1964).
45. H. Neurath, R. A. Bradshaw, L. H. Ericsson, D. R. Babbin, P. H. Petra, and K. A. Walsh, *Brookhaven Symp. Biol. 21* (1968); H. Neurath, R. A. Bradshaw, P. H. Petra, and K. A. Walsh, *Proc. Roy. Soc.* (London) (1969), in press.
46. L. Pauling and R. B. Corey, *Proc. Natl. Acad. Sci. U.S. 37*, 729 (1951).
47. L. B. Smillie, A. Furkha, N. Nagabhuskan, K. J. Stevenson, and C. O. Parkes, *Nature 218*, 343 (1968).

CHAPTER 5

1. "News and Views," *Nature 218*, 1202 (1968).
2. K. Weber, *Nature 218*, 1116 (1968).
3. D. C. Wiley and W. N. Lipscomb, *Nature 218*, 1119 (1968).
4. C. L. Sia and B. L. Horecker, *Arch. Biochem. Biophys. 133*, 186 (1968).
5. E. Penhoet, M. Kochman, R. Valentine, and W. J. Rutter, *Biochemistry 6*, 2940 (1967).
6. K. Harada and R. G. Wolfe, *J. Biol. Chem. 243*, 4131 (1968).
7. P. Strittmatter, *Ann. Rev. Biochem. 35*, 125 (1966).
8. H. C. Watson and L. J. Banaszak, *Nature 204*, 918 (1964).
9. C.-I. Bränden, *Arch. Biochem. Biophys. 112*, 215 (1965).
10. E. Zeppezauer, B.-O. Söderberg, C.-I. Bränden, Å. Åkeson, and H. Theorell, *Acta Chem. Scand. 21*, 1099 (1967).

11. L. J. Banaszak, *J. Mol. Biol.* 22, 389 (1966).
12. M. G. Rossmann, B. A. Jeffrey, P. Main, and S. Warren, *Proc. Natl. Acad. Sci. U.S.* 57, 515 (1967).
13. M. J. Adams, D. J. Haas, B. A. Jeffrey, A. McPherson, Jr., H. L. Mermall, M. G. Rossmann, R. W. Schevitz, and A. J. Wonacott, *J. Mol. Biol.* 41 159 (1969).
14. L. W. Labaw and M. G. Rossmann, in preparation.
15. E. Silverstein and P. D. Boyer, *J. Biol. Chem.* 239, 3901 (1964).
16. V. Zewe and H. J. Fromm, *Biochemistry* 4, 782 (1965).
17. W. W. Cleland, *Ann. Rev. Biochem.* 36, 77 (1967).
18. B. E. Davidson, M. Sajgó, H. F. Noller, and J. I. Harris *Nature* 216, 1181 (1967).
19. J. I. Harris and R. N. Perham, *Nature* 219, 1025 (1968).
20. O. P. Malhotra and S. A. Bernhard, *J. Biol Chem.* 243, 1243 (1968).
21. L. J. Reed and R. M. Oliver, *Brookhaven Symp. Biol.* 21 (1968).
22. H. S. Moyed and H. E. Umbarger, *Physiol. Rev.* 42, 444 (1962).
23. E. R. Stadtman, *Advan. Enzymol.* 28, 41 (1966).
24. D. E. Atkinson, *Ann. Rev. Biochem.* 35, 85 (1966).
25. J. S. Hubbard and E. R. Stadtman, *J. Bacteriol.* 93, 1045 (1967).
26. C. A. Woolfold and E. R. Stadtman, *Arch. Biochem. Biophys.* 118, 736 (1967).
27. R. C. Valentine, B. M. Shapiro, and E. R. Stadtman, *Biochemistry* 7, 2143 (1968).
28. J. C. Gerhart and A. B. Pardee, *J. Biol. Chem.* 237, 891 (1962).
29. J. Monod, J.-P. Changeux, and F. Jacob, *J. Mol. Biol.* 6, 306 (1963).
30. J. Monod, J. Wyman, and J.-P. Changeux, *J. Mol. Biol.* 12, 88 (1965).
31. J.-P. Changeux, *Sci. American* (April 1965), p. 36.
32. R. J. Watts-Tobin, *J. Mol. Biol.* 23, 305 (1967).
33. D. E. Koshland, G. Nemethy, and D. Filmer, *Biochemistry* 5, 365 (1966).
34. D. E. Koshland, *Cold Spring Harbor Symp. Quant. Biol.* 28, 473 (1963).
35. D. E. Koshland and K. E. Neet, *Ann. Rev. Biochem.* 37, 359 (1968).
36. R. R. Porter, *Sci. American* (October 1967), p. 81.
37. E. A. Kabat, *Structural Concepts in Immunology and Immunochemistry,* Holt, Rinehart and Winston, New York, 1968.
38. R. L. Humphrey, *J. Mol. Biol.* 29, 525 (1967).
39. D. J. Goldstein, R. L. Humphrey, and R. J. Poljak, *J. Mol. Biol.* 35, 247 (1968).
40. G. Rossi and A. Nisonoff, *Biochem. Biophys. Res. Commun.* 31, 914 (1968).
41. W. D. Terry, B. W. Matthews, and D. R. Davies, *Nature* 220, 239 (1968).
42. J. Almeida, B. Cinader, and A. Howatson, *J. Exptl. Med.* 118, 327 (1963).
43. A. Feinstein and A. J. Rowe, *Nature* 205, 147 (1965).
44. R. C. Valentine and M. Green, *J. Mol. Biol.* 27, 615 (1967).
45. R. C. Valentine, in *Gamma Globulins* (Third Nobel Symposium), J. Killander, ed., Wiley-Interscience, New York, 1967, p. 251.
46. H. J. Müller-Eberhard, *Advan. Immunol.* 8, 1 (1968).
47. *Complement* (Ciba Foundation Symposium), G. E. W. Wolstenholme, ed., Little, Brown, Boston, 1965.
48. W. F. Rosse, R. Dourmashkin, and J. H. Humphrey, *J. Exptl. Med.* 123, 969 (1966).
49. H. E. Huxley, *Sci. American* (November 1958), p. 67.
50. F. A. Pepe, *J. Mol. Biol.* 27, 203 and 227 (1967).
51. E. Racker, *Sci. American* (February 1968), p. 32.
52. A. L. Lehninger, in *Molecular Organization and Biological Function,* J. M. Allen, ed., Harper & Row, New York, 1967, p. 107.
53. A. Rich, in *Molecular Organization and Biological Function,* J. M. Allen, ed., Harper & Row, New York, 1967, p. 20.
54. E. Kellenberger, in *Principles of Biomolecular Organization* (Ciba Foundation Symposium), G. E. W. Wolstenholme and M. O'Connor, eds., Little, Brown, Boston, 1966, p. 192.

INDEX

NOTE: Asterisk (*) after a reference to an amino acid indicates an involvement at the active site of an enzyme. Dagger (†) indicates a coordination to a metal ion.